Good Flies

Books by John Gierach

Death, Taxes, and Leaky Waders: A John Gierach Fly-Fishing Treasury
Standing in a River Waving a Stick
Fishing Bamboo
Another Lousy Day in Paradise
Dances With Trout
Even Brook Trout Get The Blues
Where the Trout Are All As Long As Your Leg
Sex, Death and Fly-Fishing
The View from Rat Lake
Fly Fishing Small Streams
Trout Bum

Good Flies

Favorite Trout Patterns and
How They Got That Way

John Gierach

Illustrations by Barry Glickman

THE LYONS PRESS

Printed in the United States of America

10 9 8 7 6 5 4 3 2 1

Designed by Compset, Inc.

Library of Congress Cataloging-in-Publication Data

Gierach, John, 1946–
  Good flies : favorite trout patterns and how they got that way / by John Gierach.
    p.cm.
  ISBN 1-58574-139-6
  1. Flies, Artificial. 2. Fly tying. 3. Trout fishing. I. Title

SH451 .G543 2000
688.7'9124—dc21

                                    00-60251

Steady work quiets the mind.

—Wendell Barry

Contents

Acknowledgments

THANKS TO MY fly-tying friends: Ed Engle, Chris Schrantz, Roy Palm, Mike Price, and Vince Zounek, among others, for help, encouragement, and mostly for the good, idle conversation that leads to book ideas. Thanks, also, to Mike Clark and, again, Vince Zounek, for letting me paw through their libraries; and thanks for all of the above and more to A.K. Best, my friend and teacher.

Introduction

TYING OUR OWN FLIES is where many of us go off the deep end with fly fishing. That's how it happened with me, and I was even aware of it at the time. I mean, I was young and crazed, and I was sort of *looking* for the deep end.

But things didn't work out the way I expected them to—a trick that life has often played on me. I figured I'd learn the craft, master the standard flies everyone used, then move out into new territory and eventually come up with a selection of stunningly innovative, yet still traditional-looking patterns that would be mine alone and that would catch fish anytime, anywhere. (After reading a few accounts by famous tiers, it became obvious that's how it was done.) The only question was whether I'd keep these new patterns secret and just catch the hell out of fish, or take them public in an oversized book full of color photos, and become rich and famous.

What actually happened was, I became a hopeless tinkerer. I tied and fished every pattern I could find simply because it was there to try, and went through that inevitable phase in which the more complicated and difficult a fly was to tie, the more I liked it. I looked at artificial flies and caught real insects, talked to fishermen and other tiers, read anything anyone cared to publish on the subject and to this day I completely agree with every known theory of fly design.

Now, thirty years later, I have a pretty impressive collection of hooks and feathers, I've become an adequate tier almost by accident, and I've decided that many of the fine points of fly tying are lost on the fish.

From this vantage point, it seems to have taken me three decades to arrive at the obvious, but it wasn't that simple. In that time half of the best fishermen I met had museum-quality tackle and hundreds of flies so perfect they were almost too pretty to fish. But the other half fastened their reels to their rods with electrician's tape and fished a handful of really crappy flies.

When I looked into the fly boxes of tiers who were also good fishermen—something I still like to do—I saw flies that were either long and gangly or short and stubby, neat or

sloppy, trim or fat, bright or dull. They all caught fish in the right hands, and some of the best of them really did look like drowned rats.

I took up fly tying because it seemed artistic, self-reliant, frugal, and even scientific in a quaint, naturalistic sense. I decided to try and get good at it on general principles—because that's what you're supposed to do—and as my flies got a little prettier, I did catch a few more fish. But I was learning to fly-fish at the same time I was learning to tie flies, and a stealthy, accurate cast and a good drift can make more difference than a handsome fly. Still, flies seemed crucial because it was a fly that finally connected me to a fish—however occasionally, however briefly.

I do remember catching fish on some of the very first flies I tied, as clunky and flimsy as they were. They were little browns and brookies from freestone creeks and bluegills from some quarry ponds, but they were fish, and I ate a lot of them in those days.

I also remember that my first favorite dry fly was the Mosquito because you could tie the whole pattern from a single grizzly cape, and I liked the economy of that. The pattern called for a grizzly spade-hackle tail, stripped griz quill body, griz hackle-tip wings, and a griz collar hackle. It was also a buggy-looking fly that caught trout as often as not, and if you showed it to someone who didn't fly-fish, he'd say, "Damn, that looks just like a real mosquito," even though it actually didn't look anything like a mosquito.

I liked the Hare's-Ear Nymph for the same reason. Except for a wire rib, the version of it I started with was made entirely of fur from a hare's mask.

There was a moment there when things seemed simple and my entire selection of tying materials consisted of a grizzly rooster neck, a European hare's mask, some copper wire stripped from a lamp cord, and a spool of black thread, but it didn't last. I decided I should have a lot of flies, again because that's what you're supposed to want. Some really good fishermen had lots of flies, there was a small but growing industry promoting lots of flies, and stories from fishing writers and real people alike revolved around flies.

The implication in most fishing stories was that in any given situation there was the one right fly (and in nine stories out of ten, it was a fly I didn't have). If you had that fly, you'd catch fish. If not, you were screwed, and a desperate and lasting sadness descends on you when you realize you're a fisherman who's not gonna catch a fish today.

Lots of different flies began to look like insurance policies against getting skunked, but then there were still those guys with white hair sticking out from beneath dented cowboy hats who could fish a Size 12 Rio Grande King through a hatch of No. 22 Blue-Winged Olives and catch trout. They reminded me of mountain lions: I didn't see them often, but when I did they made a big impression.

I also ran into the idea of flies as art, which further complicated things. I don't mean really well-tied fishing flies, I
mean display flies tied by people who had no intention of ever
showing them to a fish. This stuff would have just been beautiful and harmless except that the line between flies to look at
and flies for fishing was never clearly drawn, so those of us
who had it bad ended up laboriously tying dry flies with extended bodies, burned wings, eyeballs, and, at their worst, little knotted legs with knees and ankles.

We probably intended to fish them, but when it came
right down to it, we usually didn't. We had too much time in-
vested in them to risk losing them, and they didn't float right
anyway.

I even fooled around with fully dressed Atlantic salmon
flies for a while—and went salmon fishing a few times to jus-
tify that. I never got anywhere near good at tying salmon flies,
but I finally managed to make two that are passable if you
don't examine them too closely. I have them framed over my
tying desk, along with four good ones by Mahlon Osmun and
two by David Sokasits. They've almost become wallpaper by
now, but I do still stop and look at them now and then. I consider them an example and a warning. They exhibit the kind

of craftsmanship I aspire to, as well as the needless extremes to which I probably shouldn't go.

I don't think all my experimental tying was a waste of time, though. It was a way of scouting out the countryside that gave me some confidence, taught me a lot about fly design and material handling, and finally showed me what kind of flies I wanted to tie by exhausting almost all the kinds I *didn't* want to tie. Even the salmon flies taught me a few tricks that have come in handy on feather-winged streamer patterns.

I also learned that it's possible to see a fly pattern in your mind that will never quite fit on an actual hook. Writers are familiar with this phenomenon, too. E.B. White once said that a book is never as good as it is just before you start writing it.

All that tying also gave me a useful take on perfection. Everyone wants to be able to tie flawless flies with every hair in place, but at the same time anything the fish don't appreciate is sort of theoretical. In the course of writing this book I had to tie near-perfect examples of all the flies as models for the illustrator, and that showed me how far I regularly stray from perfection and still catch fish.

I learned most of what I now know and use about fly tying from my old friend A.K. Best. It was nothing formal. We worked together at a fly shop for a while, and on slow afternoons—especially later on, when the shelves were half empty and the place was quietly going bankrupt—we'd sit and tie. Mostly I just watched him, but he was a good teacher, so if I asked him something, he'd explain it first one way and then another until he saw the light go on.

A.K. was the first professional tier I ever knew, and I was—and still am—damned impressed by the downright prettiness of his flies, their uniformity, and his apparently effort-less tying speed. I was even more impressed later, when I did a short stint as a professional tier myself.

A.K. taught me a lot of the nuts and bolts of it: economy of motion, how to get proper proportions by using the hook shank as a ruler, how to tie durable flies by working at the strength limit of the materials, how to tie faster not by *tying* faster but by systematically sizing, pairing, and otherwise lay-ing out the materials beforehand, and so on. He also tried to teach me that a clean desk is a happy desk, but that one never quite stuck.

By now, with his books, columns, videotapes, and work-shops, A.K. has become a guru to a lot of tiers. I was just lucky enough to get in on that early and beat the rush.

A.K. is also a practical man. He believes that flies exist to catch fish, that they should be quick and easy to make and as thoughtlessly expendable as shotgun shells are to a pheasant hunter. Sooner or later the best flies you'll ever tie will be snapped off in fish or bushes, or occasionally chewed to pieces by hungry trout. They can be pretty, and they can and should be well tied, but extra steps and needless flying buttresses should be avoided.

Something else that should probably be avoided is point-less innovation: changing perfectly good existing patterns for

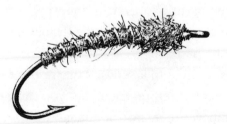

no apparent reason—except, perhaps, so you can pose as a genius. I've done this myself a time or two, so I know it's not that hard to fool yourself and others, if only because it's nearly impossible to tie a fly that won't catch a fish sooner or later.

If you diddle with fly patterns for your own amusement and that makes you happy, fine. (When it comes right down to it, there's no wrong way to tie flies.) But if it goes commercial, it can get a little insidious. I've come to believe that the sport is cluttered with flies that were tied for no other reason than to sell them to people who already have enough flies (that and the peculiarly American belief that new, weird, and different is always better than old and reliable).

Most of the so-called new patterns that you see now weren't so much invented as they were assembled from spare parts: The front end of one fly is grafted to the back end of another; or the wing material, body color, or hackling style is changed on an old pattern and it's called new. Sometimes this is done under pressure to get something eye-catching in the next catalog or magazine or to create a demand for some new tying material, and, although I don't think it happens often, I know that at least in some cases patterns that have never actually been fished have been marketed as proven fish getters.

I also think some of us tiers are too quick to claim authorship of patterns. At this late date, there's very little that's truly new in fly tying, and most patterns fall into broad types that have been around forever in one way or another. That's why you can copyright the name of a fly, but it's nearly impossible to patent a pattern.

Most flies are like folk songs: They're out there in public where anyone can interpret them, but you can't truly *own* one any more than you can own a cat, and although Ramblin'

Jack Elliot's version of "East Virginia Blues" is all but unrecognizable at times, in the end it's the same old song.

But then the difference between needless and real innovation is sometimes a hard call. Everybody thinks they can do it better, and some of them can, and sometimes the most insignificant things can make all the difference to a trout. Back in the 1930s when Lee Wulff decided to tie a Royal Coachman with hair wings and tail and rename it the Royal Wulff, it might have seemed like a small thing, but it made the fly more durable, more buoyant and visible, and has probably kept the old pattern alive well past what would have been its prime.

I know tiers who are endlessly fiddling with patterns—old and new—for all kinds of good reasons. Some, including lots of guides, want to make them quicker, cheaper, more visible, and more durable because, as any guide will tell you, most clients can't see, and they're hell on flies.

Some tiers are in love with the lusciousness of the materials and the blank canvas of a bare hook. They're having fun, they're not hurting anyone, and they're probably catching fish, so they should be left alone.

Others are trying to imitate specific insects in order to catch particularly cagey trout, or trying to come up with something just a little bit different for fish that have already seen everything that's for sale at the local fly shop.

Still others are trying to clarify their thinking about why a trout eats a fly in the first place, a question over which lives have been wasted. In the end, the beauty of fly tying is that there are all kinds of ways of making it work, as well as all kinds of ways of caring whether it works or not.

Still, now and then someone does come up with something the fish can't leave alone. It might be a little bit different than the flies you're used to but it still has that recognizable look. As Tom McGuane said of the Adams dry fly, it "says 'bug' to the trout in a general yet friendly and duplicitous way." When I see a fly that does that, I politely ask to borrow it, and I study it for a minute so that maybe I'll be able to reproduce it back home after I've lost it to a pine tree, all the time hoping I haven't given this particular tier the "pointless innovation" lecture recently.

I guess I decided to write this book now because my fly tying has finally settled down to match the way I fish. Don't get me wrong, my fly boxes are still a mess, I still gratefully accept strange flies that work from friends and kindly strangers, I still have a weak spot for odd, local fly patterns, and I still carry more flies than I need. But I've also arrived at a core handful of patterns that I tie the same way year after year, that I usually either start with or come back to, and that in fact catch most of my fish.

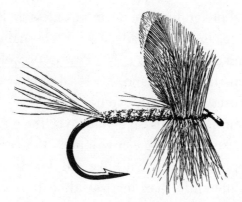

I think of these flies as regional to a stretch of the Rocky Mountain West, roughly from Colorado to Alberta and British Columbia, because that's where I do most of my fishing, but a few of them have also worked well in exotic and faraway places like Labrador and Pennsylvania. They reflect an inordinate fondness for mayflies and a lot of freestone stream and lake fishing, but some are pretty much meant for tailwaters and spring creeks, where things can get a little tricky.

I think most of these flies are pretty in a plain, homespun sort of way, but a few of them might seem a little on the fancy side to a real simplicity freak. On those, I've either decided that the extra features aren't that hard to tie and really add to the effectiveness of the fly, or I just thought they were too sexy to resist.

There's a fine line between practicality and snazziness in flies, especially when you consider that you'll fish a pretty fly better than you will a homely one (never mind that a butt-ugly fly that works can start to look okay after a while), and that the best flies tend to be simultaneously workmanlike and poetic. Still, I prefer flies that are more or less traditional, that don't use rare, exotic, endangered, or unusually expensive materials, and that are fairly easy for me to tie well in fishable numbers. On the other hand, I'll break any or all of those rules for a pattern that really catches fish.

So, as it turns out, this isn't going to be the big book full of my own patterns that I once envisioned. It's a little book full of other people's patterns that I like and that, in some cases, I tie a little differently than most—although probably not as differently as I think. Whenever I invent something new, all I have to do is look in the four or five most recent fly-tying books (or the four or five oldest ones I can find), and there it will be, with step-by-step directions.

Or maybe I'll mention it to another tier, who'll say, "Yeah, I tried it like that once. It ain't worth the trouble."

Chapter 1

Materials, Hooks, and Tools

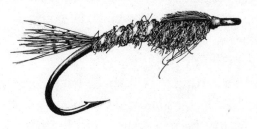

Materials

When I started tying, I didn't realize how much time I'd have to spend looking for good material; dry-fly hackle, mostly, but other things, too. One of the early lessons you learn is that skill can take you a long way, but ultimately the best you can do with mediocre materials is tie mediocre flies.

This really came home to me when I broke down and bought my first genetically raised rooster cape, which back in the 1970s would have been a Metz. It was three or four times more expensive than even the best of the Indian gamecock necks I was used to, but it was also three or four times better in every way: The colors were clearer; the hackles were longer, less webby, and came in more sizes—including No. 18, 20, and smaller—and there were more feathers on the skin. It was the beginning of a real dependence: This hackle was so good compared to what I'd been using that, like most other tiers, I immediately couldn't do without it.

Now genetic dry-fly hackle from various companies is about all you can find in most fly shops. For the most part, the quality is better now than it's ever been, but the one flaw is that the spade hackles have been all but bred out of these necks.

Spade hackles are the short, wide throat feathers that were once found at the edges of the widest part of a skinned cape. They're too stubby and wide to be used as wound hackle on anything but spider patterns, but the barbs are long, shiny, and stiff. They're the best, if not the *only* material for tails on many standard dry-fly patterns.

For a while getting decent spade hackle was a struggle. I use genetic hackle for almost all of my dry flies—Hoffman and Whiting, for the most part—and I'm happy to have it, but for quite a while I was reduced to mining old picked-over necks for spade hackle and buying second- and third-rate so-called dry-fly necks that were just about useless except for some good tailing feathers.

Things were starting to look pretty desperate, when Whiting Farms began offering Coq de Leon rooster necks and saddles with some great spade-quality hackle on them. At first it was a rumor; then small, sample patches of saddle hackle began circulating among the intelligentsia, and finally full skins began to turn up over the counter.

Tom Whiting at Whiting Farms recently told me that although the birds are being raised primarily for their spade hackles, some tiers are finding other uses for them as well. A tier I talked to in a fly shop not long ago told me that Coq de Leon saddles make great Matuka-style streamer wings.

Anyway, this stuff is magnificent, as good as anything I have ever found on a cape. The saddle hackles are lightly speckled, while the spade hackles on the necks are more of a solid color (I'm not sure which I like best). Both are long, straight, shiny, and stiff: ideal for tailing. They come in three natural colors—white, light ginger, and dun—and the white ones will be perfect for dying to other colors.

I immediately bought two saddles—a ginger and a dun— thus solving one of life's central problems with a single trip to the fly shop.

It had been a close call, because at the height of the shortage of spade hackle John Betts started marketing eight different colors of plastic artist's brush fibers as tailing material. I understand that this stuff works well and a lot of tiers like it, but I have never tried it. I have managed to resist cell phones and e-mail, and I didn't think I could come to terms with plastic hackle either, although if Whiting hadn't saved me at the last minute, there's no telling what I might have done.

I do try to stay away from synthetic fly-tying materials on general principles, but it's not something I'm willing to fight about. I just like the natural stuff. I like the look, the feel, and even the smell of it. I like the way flies look when they're tied from real fur, feathers, and hair, and I even enjoy all the pawing and snooping and bartering it takes to get the really good stuff. (Not long ago my friend Vince Zounek and I concluded an elaborate, year-long series of trades involving, among other things, artwork, a Kentucky rifle, boxes of hooks, and some good dry-fly necks. At one point I said, "You're keeping track of this, right?" He said, "I thought *you* were keeping track of it.")

Anyway, my rule of thumb is that I won't use a synthetic if there's a natural material that will do the job, but I'm not a complete Luddite, and I can tie the occasional wisp of poly yarn into a midge pupa without feeling like I'm doing the devil's work.

If there's one drawback to natural materials, it's moths, and the more material you have, the more likely you are to get them. All I know is that you have to keep the stuff clean and store it in sealed containers with moth balls. If the room where you keep your tying stuff doesn't smell like your grandma's closet, you are courting disaster.

Natural materials vary a lot in quality and the premium stuff is almost always scarce, so I've been squirreling away good materials for years. I concentrate on the material I use a lot and know I'll need, but I've also been known to packrat things on general principles. I have no use whatsoever for javelina hair, for instance, but I do have half a prime, tanned skin from

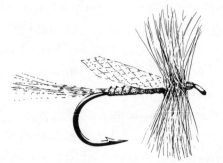

Texas. Maybe I'll need it someday, or, better yet, maybe some-one with a stash of something I *do* need will want it.

I also have a cigar box full of old Indian gamecock necks, all grade-As in furnace brown, badger, ginger, and cream vari-ant. (Blue-dun gamecock has always been as rare as hen's teeth.) These little necks have good dry-fly hackle in Sizes 16 and bigger, and sometimes it's fun to show them to younger tiers and say, "Yup, back before Metz, Hoffman, and the rest, this was about the best you could get."

I tie a lot of dry flies with hen-hackle wings, and good hen necks are hard to come by. The perfect dry-fly hackle would be long and thin, with all barbs and no web, but the perfect hen hackle for winging would be just the opposite: short and round and *all* web. Most are about half and half, without enough web to make a clear silhouette of the wing. By system-atically going through every neck in every fly shop I go into and by now and then getting in on bulk shipments to profes-sional tiers I know, I have managed to beg and high-grade a supply. Still, keeping myself in these feathers is a matter of tireless vigilance.

The best hen hackle for dry-fly wings tends to come from run-of-the-mill barnyard chickens. The genetic hen hackles I've seen are too long and pointed and have too much barb for wings, but Whiting Farms is now raising some small chickens especially for small hackle-tip wings that I've been told are very promising. They're calling them Mayfly Wingers.

I use hen necks for hackle tip wings on small flies because I've yet to find any other material that comes dependably in the right sizes, but the same style wing on larger flies gives you a little more leeway. You can find beautiful, webby, teardrop-shaped gray feathers on some dyed Indian, Chinese, and domestic hen backs and on the breasts of blue grouse, although even the smallest of these feathers are usually too big for a Size 16 fly. Some genetic hen *backs*—not necks—also have good, larger-sized webby feathers for wings. The shoulder feathers from some domestic pigeons are also good and will sometimes work for wings down to a Size 16.

The turkey T-base feathers that make such great wing posts on parachute-style dry flies present the same problem as hen hackle. Many suppliers sell turkey back feathers, called flats, for wing post material, but those feathers have a fringe of fine barbs along the top and don't make a good, solid-looking post. T-base feathers grow on the necks and breasts of turkeys, and they're all web from stem to tip. T-base can be harder to find than flats, if only because many fly shops don't know the difference. Right now I have two ounces of prime, hand-picked T-base stashed away—almost enough to fill a throw pillow. If and when more comes along, I'll grab it.

I like Hungarian partridge flanks and back feathers for nymph legs and wet-fly hackle—and I have a good supply of skins gleaned from bird-hunting friends in Montana—but unfortunately that stuff doesn't come in sizes small enough for flies much under Size 14. For flies smaller than that I usually use Indian hen necks, which come in various shades of brown, honey, and dark ginger, *if* you can find them.

Most of my Indian hen necks date back a dozen years or so, and I don't see them for sale much these days. I'm not too worried, though, because in a pinch I think a poor-quality domestic hen neck dyed a medium honey brown would work just fine. The only problem I foresee is that I'd probably ruin a neck or two trying to get the right dye bath, but it wouldn't be the first time.

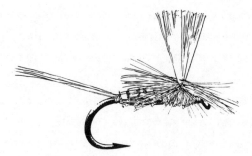

Now and then I'll use some other soft hackles for small nymphs and wets. The small, mottled, grayish to brownish cape feathers from quail and various grouse can be good, and sometimes there's nice soft hackle on the shoulders of the wings. I skin most of the game birds I shoot and beg skins from other hunters when I can. Sometimes when I'm tying a batch of small flies I'll sort through the box of odd skins looking for just the right color and mottling pattern—more to please myself than the fish.

I've also developed a taste for white-tailed ptarmigan flank feathers. Ten years ago I kept the feathers from the first few ptarmigan I shot during the short, three-week hunting season here in Colorado, and they really dressed up my next batch of golden stones and soft hackles. The flank feathers are as soft as those of partridge, and they're a warm, yellowish, golden brown color with very fine chocolate-colored speckling. A ptarmigan also just happens to be the best-tasting game bird I've ever eaten, so now, every September, some friends and I make the god-awful slog up to twelve thousand feet to collect ptarmigan hackle and that one great meal.

Good paired oak-colored wild turkey wing feathers—for the wings on Muddler Minnows—are sometimes hard to come across, so I high-grade those when I find them. Size, color, and mottling pattern are considerations; another is that some so-called paired feathers come from different places on the opposing wings, and in some cases, they're not even from the same bird.

The same goes for wood duck flank and bronze mallard, for the trailing husks on emerger patterns, as well as the nice flat turkey marabou blood feathers that make such neat Woolly Bugger tails: When I find really high-quality examples of any of these, I buy them and stash them.

A few years ago I stumbled onto some great peacock herl for small flies. The peacock herl you get commercially for fly tying comes from the Indian blue peacock and has those huge, fully eyed tail plumes. But there's a smaller bird native to Africa, called a green peacock, that has shorter, less spectacular tail feathers with much finer herl. The smallest of these feathers are proportioned about right for the tiniest flies you're likely to tie—Sizes 20 on down to 24 or 26.

I've never seen green peacock for sale anywhere at any price, but I managed to salvage the feathers of a local pet bird that met an untimely end. (I didn't even know what it was until a bird expert told me.) I gave a few feathers away to friends, but I hoarded what should be a lifetime supply as long as I don't get desperate and start tying professionally again.

I'm sort of a fanatic for Hare's-Ear Nymph patterns—I regularly fish several different ones—so I'll usually buy hare's masks by the half dozen and blend up a gob of rough dubbing the size of a grapefruit. I like my hare's ear dubbing to be on the dark side, so I just blend the darker hairs between the ears and down the middle of the mask where the underfur shades from gray to dark tan with black and brown tips. The lighter, gray to beige to cream fur I blend separately for lighter-colored patterns.

I've settled on rabbit underfur for dry-fly dubbing, and I typically buy whole dyed or natural skins. Sometimes I blend the colors to get just the right shade, and sometimes I take the fur straight off the skin for things like the plain yellow I use for Pale Morning Dun dries.

It's a little difficult to prepare dubbing from whole skins because the stiff guard hairs that you don't want in the blend for dry flies go all the way down to the skin and have to be plucked out; but once that's done, you just shave the soft underfur off the skin into the blender, and you get a fine, soft, natural, easy-to-work-with dubbing that can go on heavy if that's what you want, or so light it doesn't do much more than color the thread.

Speaking of thread, I'm pretty easy to please there. On most patterns I like the thread to roughly match the color of the body—purely for looks—but it doesn't have to be perfect. I also like to use the finest thread I can get unless I'm tying streamers or big, clunky nymphs that need bulk. Fine thread lets you take more wraps for durability and allows you to keep your heads small and neat. Right now I'm using Giorgio Benecchi's Ultrafine, Gordon Griffith's Ultrafine, and Orvis 8/0 more or less interchangeably.

A.K. long ago convinced me of the importance of quill-bodied dry flies. They're realistically trim, and they have the

hard, not-quite-shiny look of a real bug body along with a subtle, natural-looking segmentation. I like stripped quills, dyed or natural. They're easy to work with, durable on the fly, and they usually won't crack or split if you soak them in water for a few minutes before tying.

For some flies I use goose biot because it lies nice and flat on the hook shank, and when you wrap it in the opposite direction of its slight natural curve, it has a dark, slightly fuzzy trailing edge that acts as a rib. If you want the same effect on larger flies, use turkey biot. As with stripped quills, they work better if you soak them first.

Stripped, dyed quills and biots are available commercially, but sometimes the colors described in the catalogs aren't quite what you expected, and now and then you'll get a batch that was fried in the bleaching and dying process and is too brittle to use. Still, I'd rather buy the stuff than go through the laborious process of stripping and dying myself. I *have* done some of my own dying in the past and it has usually worked out well enough, but it's a pretty serious production and I'd rather spend that kind of time and effort tying or fishing.

When I'm reduced to doing my own dying, I always consult A.K.'s definitive book, *Dying and Bleaching Natural Fly-Tying Materials*. (Note that that's "natural" materials.) A.K. has reduced this to a science, and, although I haven't given color nearly the kind of thought he has, his colors just look right to me. If you follow his directions to the letter, things will likely come out right.

With the exception of thread, floss, hooks, and a few other odds and ends, I like to buy my material hands-on rather than through the mail so I can look at it first. That's especially important when I'm looking for dyed colors or very specific natural ones. I've seen an almost infinite variety of green to gray materials that were called "olive," and not long ago I got into a vicious cycle of returning cream necks to a supplier, each time saying that what I wanted was *ginger*. Just the other day I saw some dark dun dry-fly necks that I'd have called black, as well as some dyed dun hen necks that had a sickly purple cast to them.

Quality can be pretty subjective, so that one tier's Grade AA Premium can be another's Just Barely Usable. I'm always happiest when I can dig through a big pile of material looking for the piece that's just right. It's best when I'm not desperate, so if none of it is really good, I can go somewhere else and dig through another pile.

Hooks

When I started out, it was generally thought that the best fly-tying hooks were Mustads, and that made a kind of intuitive sense. They were made in Norway, and who would know more about fishing than Norwegians? I still use Mustads for almost all of my dry flies, even in the face of all the other choices we have now. At this stage of the game, the dry flies I tie just fit on Mustad hooks, and they've worked so well for me for so long that I've just never seen a reason to change.

For a while there was a rumor going around that Mustad hooks were brittle and broke easily, something I haven't noticed in thirty years. I heard this so often around Colorado that I finally asked one of the guys in a fly shop if *he* had ever had Mustad hooks break.

"Well, no," he said.

"Then who told you they break?" I asked.

He said, "The Tiemco rep."

I like the Mustad 94840 standard length, standard wire hook for dry flies and some nymphs, especially in Sizes 16, 18, and smaller. I prefer the 94831, which is the same gauge wire in a 2X long shank for some larger mayflies and caddis, as well as hoppers. I especially like the way this hook makes a longer abdomen on mayfly patterns. The only problem I have with the 94831 is that it's made only down to a Size 16. I'd really like to see it in Sizes 18 and 20 for small mayflies.

I use the Tiemco 200R hook for lots of streamers, nymphs, and emergers. It's a straight-eyed, strong-wire, 3X long hook with what they call a semidropped point, which means that on most flies the base of the tail will be more or less above the hook point. I started tying on them because they're an elegant shape and they make a graceful-looking fly, but they also turned out to be sharp and durable and they hold fish well. I also like them because they come in a full range of freshwater sizes, from 4 to 22.

Another hook I use for streamers and nymphs is the Dai-ichi 1870 Swimming Larva hook. It looks a little like a long, stretched-out English bait hook with a hump in the shank and a little dip in the wire right behind the straight eye. I started fooling around with them when they first came out a

few years ago and found that if you tie dumbbell eyes in that dip behind the eye, you'll get a fly with lead eyes that still swims with the hook bend down. (Weighted eyeballs tied on top of a hook shank usually counteract the weight of the bend, turn the fly over, and make it swim upside down, which is fine if that's what you want it to do.)

I asked Bill Chase at Daiichi if the 1870 was intentionally designed so it would swim right side up with lead eyes tied on top. He said no, that it just turned out to be an "interesting property" of a hook designed so that the eye was lower than the point.

For my purposes, the hump in the shank is a little too severe on the smaller-sized hooks, so I bend them open a bit before tying. They come in sizes 4 through 14, but I think they'd be useful both larger and smaller. Like the 200R, these are strong, graceful hooks that make pretty flies.

Those are the hooks I use for most of my flies, but I use a few others from time to time. I tie my Tarcher Nymphs on English bait hooks (Mustad 37160, sizes 16 through 10) because that's how Ken Iwamasa designed the fly. For some large streamers in Sizes 2 and 4, I'll sometimes use a Mustad 38941, and for some classic old streamer patterns I'll use a long-shanked, Limerick bend hook like the Mustad 2665A, just because that old-timey shape hooks and holds fish well and looks so right on some traditional patterns.

For upside-down streamers and anything else that requires a straight-eyed streamer hook, I'll use the Mustad 9674, the Daiichi 1750, or the Tiemco 9395. For a water boatman pattern I use a short-shanked scud hook like a Mustad 80250 or a Tiemco 2457.

When it comes right down to it, I guess I choose hooks as much for how they make the flies look as for their performance, and I probably pick most of my other materials for the same reason. I'm an unreconstructed romantic when it comes to good-looking flies. They make me happy, and I think they catch more fish, although I couldn't begin to tell you why or explain why ratty-looking flies catch fish, too.

I haven't done extensive testing of hooks—as you can tell—but my guess is that they all work well enough for day-to-day fishing. At least that's what I seem to hear from tiers who are stuck on one or another brand or style. Pick any weird, obscure, specialized hook and you can find a fisherman somewhere who thinks it's the best thing since sliced bread. For my own part, I can't think of a single fish I've lost, large or small, that I could honestly blame on the hook. A few were from bad tippet material; all the rest were from operator error.

Tools

I think it's safe to say that the burning question in modern American fly tying is, Do you really need a $500 vise? The answer is no. You may *want* a $500 vise, for reasons of your own, but all you really need is something that will hold the hook still long enough for you to tie a fly on it.

I wish I could rate the vises on the market now, but the fact is I haven't seriously looked at tying vises for a long time, although some of the new ones I've seen have heads cocked at uncomfortable-looking angles and way too many knobs and levers.

For fifteen years or so, I did all my tying on an old L&H vise. (No one I know who's under fifty has ever seen one before, so I have to assume it's no longer made.) This is a beautifully designed tool that's durable, easy to use, and has precisely five moving parts, including the C-clamp. It holds

hooks from Size 4 on down, and it's stood up to fifteen years' worth of hard use. It probably cost around $25 in the late '70s or early '80s. I don't know if there's anything comparable on the market now, but I sincerely hope there is.

A few years ago I bought a used HMH, just because I always wanted one, and I now use that for most of my tying, although I still bring out the old L&H for big nymphs and streamers. Beyond that, I've essentially used the same few fly-tying tools for the past fifteen or twenty years with no complaints. I like Wiss scissors for the neat way they nestle in the palm of my hand, leaving my fingers free to handle materials and saving me from having to put the scissors down and pick them up a dozen times in the course of tying a single fly. The blades come nice and sharp with even tips, and they're replaceable, so when the old ones finally do start to get dull, you can put on fresh ones. I bought mine at a fabric shop back when they were mostly sold to sewers. I still use that first pair, although I've probably gone through a dozen sets of blades.

I can think of only two tests for fly tying scissors: They should be comfortable to hold, and they should cut cleanly right down to the tips.

I use the same English-style hackle pliers I started tying with. They're medium sized and made of strong wire, and they have smallish, rounded jaws that fit together nice and flat. When I got them, one jaw hung out over the other slightly and they had a tendency to cut the hackle they were holding, but I filed them and smoothed them with emery paper, and they've been fine ever since. They'll last a lifetime if I don't lose them.

I use the old-style Thompson whip finisher. I'm told there are now whip finishers on the market that are easier to use, but it took me months to learn how to operate mine, it's become second nature, and I don't see any reason to abandon a hard-won skill.

I have two sets of hair stackers, in sizes regular and magnum, one machined by an old friend years ago, the other commercial. They're both the old one-tube-inside-another style, where you load the hair in the inside tube and tap it on the desk to even the tips. (I once had a dog who was so stupid that he always thought someone was knocking on the door whenever I tapped the tube. I must have shown him what I was doing a hundred times, but he never got it.) I use the two normal-sized ones interchangeably, and I never use the big ones because I never use that much hair at one time.

I have an odd assortment of thread bobbins, mostly standard Matarelli's but also a few others of that same wire frame and bead design. Most have short, fine tubes, but a couple have the long, flared tubes that would make them "floss bobbins." I use whichever one is handy at the moment. I have all of them bent open enough so that there's almost no drag on the thread. That way I can apply tension by palming the spool.

I also have a few homemade bodkins, a pair of cheap, clunky scissors for cutting copper wire and lead, and a pair of small wire cutters for nipping bead chain.

The only new tool I've added and stuck with in recent memory is a $2.95 wire dubbing twister. I don't use it often, but when I need it for hair hackles or spun-fur streamer heads, it's a lifesaver.

I have tried a few other gadgets over the years, but none of them seemed useful. Anyway, as someone once said, your best fly-tying tools are your eyes and your fingers. If there's a secret to tools, it's that a few good ones make things easier, but too many just get in the way.

Aside from some basic tools, you will need adequate light that shines brightly on the fly but not into your eyes. I use a cheap hooded, goose-neck office lamp with a 60-watt bulb, but of course you can get something fancier if you like.

You'll also need sharp vision. I once tied with my naked eyes, but I now use 2.5-power over-the-counter reading glasses, and take comfort in the knowledge that those things are available all the way up to 5-power. I once tried a magnifying glass, but I didn't like it, although other tiers swear by the things.

Some tiers get downright temperamental about their desks and chairs, but I think all you need is a desk or table at a height that doesn't make you hunch uncomfortably over the vise and a chair you can sit in for a long time without having your butt go to sleep.

I will say that efficient material storage and a neat, well-organized desk are a big help, but I have to admit that I don't always follow that advice to the letter. Now and then I have to search for half an hour for a little snippet of something that I didn't put away in the right box the last time I used it. And when I'm tying fill-ins (two or three each of several different patterns) I sometimes end up with the materials for half a dozen different flies on the desk at the same time, making a pile of stuff under which I can loose my scissors.

Maybe it's best to say that you should get as organized as your character allows you to be, and then just tie your flies.

Chapter 2

Small Mayflies

For as long as I've fished with a fly rod, I've had a self-conscious weakness for dry flies; first because of their puffed-up classiness, later in spite of it. Dry-fly fishing may or may not be the most demanding way to catch fish, but everything about it is visual and beautiful and I've always been a sucker for that kind of thing.

I also think it must be what fly fishing was all about in the beginning. The earliest accounts of fishing with artificial flies are vague, but they seem to describe dry flies if that's how you choose to read them, and most of the oldest patterns you can find at least have wings. When you think about it, the original idea of fly fishing—not to mention the name—had to come from fish feeding on floating insects. You'd see trout making rings on the surface, you'd see flies disappearing in those rings, and you'd quickly learn that a real fly impaled on a hook wouldn't float. You'd get to thinking about this, maybe even lose sleep over it, being a fisherman and all.

Eventually, dry-fly fishing became annoyingly high tone: the only proper way for a gentleman to fish in places like the English chalk streams of the late 1800s. When G.E.M. Skues first started writing about upstream wet-fly and nymph fishing, he had to carefully explain that these were "minor tactics" that were not meant to "supplant or rival" dry-fly fishing, presumably for fear of being excommunicated from the Fly Fisher's Club for heresy.

That should seem ridiculous to us now, but oddly it doesn't, probably because American fly fishing has such deep English roots. We still chase hatches for the dry-fly fishing; more amateur entomology is inflicted on floating mayfly patterns than on any other kind of fly; and every once in a while someone still puts himself up on a pedestal by claiming to be exclusively a dry-fly man (not seeming to realize that a lot of younger fly casters can't see the pedestal). He probably does it because he likes the poetry and the prettiness and that patina of Old World snobbery, which is, in fact, about the same brownish amber color as a hundred-year-old bamboo fly rod.

Blue-Winged Olive

I don't know if the Blue-Winged Olive mayfly is my favorite hatch, but it's definitely the one I've fished the most. There are mayflies you could call Blue-Winged Olives in just about every western tailwater, spring creek, and freestone stream I know of. They range in size from a 16—which is a big one—down to about as small as they make hooks.

In their book *Hatches*, Al Caucci and Bob Nastasi say that the name Blue-Winged Olive "has been indiscriminately applied to over twenty mayfly species," and that this has caused

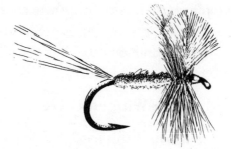

Blue-Winged Olive

a lot of confusion. Honestly, I haven't noticed much confusion. Most of the fishermen I know don't worry about the Latin names for bugs; they're just happy that one fly works on so many different hatches.

I usually even use a Blue-Winged Olive pattern when I'm fishing a *Callibaetis* hatch on a trout lake. For years I tied these flies with speckled wings made of gray partridge, teal or mallard flank to try and match the natural Speckled Duns, but I finally realized that most days the fish were just as happy with an Olive pattern with a plain gray or even a white wing. It was one of the few times I was actually able to do away with a fly pattern in the interest of efficiency, which I suppose is a good thing.

I've been staying away from the big, crowded tailwaters more and more over the past few seasons, but I used to haunt the Blue-Winged Olive hatches on the South Platte River spring and fall. There were at least three sizes of insect, a No. 18 or 20 that everyone called a Baetis, a No. 22 or 24, and sometimes a tiny No. 26 or 28.

To me, the bugs looked identical except for their size—grayish olive bodies with gray wings—so I think of the patterns as interchangeable, although there are tiers who will argue with that. I tie most of my Olives in Sizes 18 and 20, but I like to have a few 16s and some 22s and 24s, just in case.

Dubbed-Body Blue-Winged Olive

When I first started fishing this hatch—going on thirty years ago—a standard Blue Quill, a dubbed-bodied Blue Dun or even an Adams in something close to the right size worked well enough (and still does in most places), but I soon settled on the Blue-Winged Olive pattern A.K. was using at the time. It had a dun spade-hackle tail, dun hen-hackle wings, dun collar hackle, and a trim body of A.K.'s special dubbing mix: a pale grayish olive blended from five different colors of dyed rabbit fur. The body color was lighter than most Blue-Winged Olive patterns (most were, and still are, too green and too dark), and the wings and tail were both slightly longer than on the proportion charts that were in just about every fly-tying book published at the time.

Somewhat oversized wings and long tails cocked slightly upward were a hallmark of A.K.'s dry flies then—less so now that so many other tiers have picked up on it—but at the time there were those who said the flies were tied incorrectly. To that, A.K. would ask rhetorically, "How can they be wrong if the proportions match the bug instead of the charts?"

Olive Dun Quill and Parachute

When A.K. started tying what he called the Olive Dun Quill, which was that same pattern except with a dyed olive quill body, we all jumped on it. It was real pretty in the best way: a

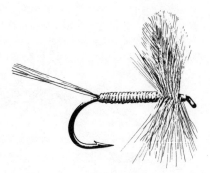

Olive Dun Quill

trim little quill-bodied dry fly that could have been tied a century ago except that it had that realistic grayish olive color to the abdomen. It was easy to tie, it caught fish, and in the first few seasons before the pattern became popular, there couldn't have been a half dozen of us fishing it, so most of the fish had never seen one.

There was also a parachute version with a white turkey T-base wing and parachute hackle, and one or the other of those two quill-bodied patterns is still what I start with on any Blue-Winged Olive hatch I fish.

When neither of those quite gets it, I still carry the dubbed versions of both flies, and although most of my Olive parachutes have the standard white wing post for visibility, some have pale gray wings for realism and a few have a cut wing made from a plastic winging material called Zing. All of them are tied on Mustad 94840 hooks in Sizes 16 or 18 down to as small as I can manage them.

You're right, all this does sound a little compulsive, but slight variations can often make a difference in whether a fly works. No matter how I normally tie a fly, I often tie a few

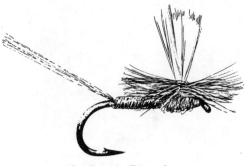

Olive Dun Quill Parachute

another way—say, with dun-colored wings instead of white ones—and sometimes it makes a difference.

Not that any of them always work. No fly *always* works, especially on a heavily pounded catch-and-release fishery like the South Platte, where the trout have gotten harder and harder to fool over the years.

There were autumns on that river when the Blue-Winged Olive hatches stretched on for six weeks and one gloriously mild year it started in September and lasted until late November. A bunch of us got into a jag of driving down there twice a week, and it seemed like on every trip the water would be lower and clearer, the bugs would be smaller, and the trout would be smarter.

In the beginning we'd just whack fish, but by the end of it we'd be right at the edge of possibility: There would be ice forming in our guides, a fire going on the bank, and we could cast all day for a couple of strikes. Eventually we'd just get blanked, and I guess that was the point. I think we wanted to experience the exact moment when it became impossible to catch a fish and it was too damned cold to be out anyway.

A Blue-Winged Olive hatch is the one that's most likely to reduce me to a sheer panic of fly changing, if only because I carry so many patterns for it and fish it so often on technical tailwaters and spring creeks. I try to resist changing flies because I know in my heart that I should try everything else before I switch. I mean, maybe I'm not getting strikes because it really is the wrong pattern, but it's just as likely to be a poor presentation, a bad drift, a leader that won't allow the fly to bobble naturally on the current, or any of a dozen other microscopic details. I think some of us carry too many flies simply because we don't give any one pattern a fair chance.

Blue-Winged Olive No-Hackle

On days when none of the conventional Olive Duns will work—and I've satisfied myself that I'm not doing anything wrong—I'll sometimes do okay with a No-Hackle. No-Hackle mayflies were popularized by Doug Swisher and Carl Richards in the 1970s. Their favorite, and the one you saw tied commercially, was the Sidewinder, with a forked tail of hackle fibers, a dubbed body, and duck-quill segment wings tied off the sides of the hook shank. It was tied in sizes and colors to match different hatches, and of course there was no hackle, hence the name.

Blue-Winged Olive No-Hackle

This is a good-looking and effective fly that works well in slow, smooth currents, but it's difficult for me to tie because I'm all thumbs with quill wings. It's not a very durable pattern, either. Quill-segment wings are delicate even when they're sprayed with a fixative, and after a few fish they're frayed so badly they might as well be made from dog hair.

Maybe I wasn't the only one who had trouble, because in *Selective Trout*, published in 1971 (and to be republished this year by The Lyons Press), Swisher and Richards suggested some other dressings for the No-Hackle, with wings tied from hen hackle, deer hair, elk hair, and duck shoulder feathers.

I like the duck shoulder pattern because it's quick and easy to tie, it looks right, it's durable as hell, and it's deeply obscure. Even the Sidewinder has been dropping out of fashion, and the duck shoulder No-Hackle was never popular. In fact, the ones I tie are the only ones I've ever seen outside of that book, and sometimes the way to catch an especially difficult fish is to show him something the likes of which he's never seen, but that still sort of looks like the bugs he's eating.

This really is an elegantly simple fly pattern. The split tails act as outriggers to keep the fly floating upright, and the wings are set slightly farther back on the hook shank than most, with the body dubbed around and ahead of them, so its proportions are more like a real mayfly dun than most standard patterns. I like to make the thorax slightly fatter than the abdomen, which is also more realistic.

I tie these in two ways: with a body dubbed from A.K.'s olive dubbing blend, and with a body of either olive-dyed quill or light olive goose biot with a dubbed thorax. I usually use goose shoulder instead of duck shoulder because the feathers are lighter gray (more like real mayfly wings, and also easier to see), and because I have some goose-hunting friends who give them to me.

I do better with the quill-bodied fly, probably because I think it's the prettiest, so I always start with it, but any kind of No-Hackle dry fly is a specialized pattern, good only for slow, smooth currents where it will stay afloat, so I don't use them very often.

Blue-Winged Olive Palm Emerger

The Blue-Winged Olive emerger pattern I like came from Roy Palm on the Frying Pan River years ago. When I first saw one, it didn't look like much, but even then I knew enough to try any fly Roy recommended, no matter what it looked like.

I tie it on either a Tiemco 200R or a Mustad 94840 hook in sizes 18 and 20. It has a long, sparse tail of wood-duck flank for a trailing nymph shuck, a body of olive goose biot, a small, dubbed thorax (I use A.K.'s Blue-Winged Olive blend), and a sparse, dun hen collar hackle. It looks like a traditional soft-hackled wet fly, and you can squeeze it wet so it sinks a

Blue-Winged Olive Palm Emerger

fraction of an inch or grease it so it floats low in the surface film, where it works as an emerger or a crippled dun. Either way, it's a little suggestive wisp of nothing on the water, which is sometimes exactly what fish want when they won't rise to a proper-winged dry fly the way they're supposed to.

RS-2

Another fly I use as an Olive emerger is the good old RS-2, or Rim, Style Two, first tied by Rim Chung of Denver for the Blue-Winged Olive hatches on the South Platte River. It has a tail of two split beaver guard hairs or moose body hairs, a dubbed fur body, and a stubby wing made from a clump of gray down from the base of a gamebird feather: pheasant, grouse, partridge, or whatever. I tie it on a 200R in 16, 18, 20, and 22 in the original muskrat gray and in pale olive, and I like to set the wing in the thorax position and dub ahead of it to finish the body. Many tiers just tie the wing in at the head, but, as Roger Hill says in his book *Fly Fishing the South Platte River,* exactly where the wing goes on this pattern is "a matter of considerable indifference to the trout."

RS-2

Pale Morning Dun

I don't carry as many different flies for the Pale Morning Dun hatch, but that's only because I don't fish it as much as I do the Olives and so haven't been driven crazy by it quite so often. Still, I'm prepared.

I tie my Pale Morning Duns on Mustad 94840 hooks in Sizes 16, 18, and 20 to match the sizes of the small yellow mayflies I see most often in the West. Actual Size 20s seem rare, although I've seen tiny ones on the Henry's Fork in Idaho and a few spring creeks, and I've met fishermen who tie these as small as a No. 22. Some of these guys will tell you that the 18s are the *Ephemerella infrequens* and that the 20s and 22s are the *E. inermis*. Others refer to them as the little yellow ones and the *real* little yellow ones.

Now and then, though, a Size 20 will work better for me than a more life-sized 16 or 18. I have no idea why that should be, but it's something Ed Engle turned me on to years ago: the idea that a snooty trout that won't eat a Size 18 fly may well eat the same pattern tied on a No. 20 hook, for reasons known only to the fish.

Frying Pan Pale Morning Dun

One of the best flies for this hatch is A.K.'s Frying Pan Pale Morning Dun. It's an easy enough fly to tie, but it's hard to make a proper one because every piece of material on it has to be dyed to just the right color. The body is dyed pink quill, the hen-hackle wings are dyed to a color A.K. would describe as a pale creamy tannish gray, and the collar hackle and spade-hackle tail are both dyed a very specific golden color with an almost but not quite metallic sheen to it.

Frying Pan Pale Morning Dun

This is one of those flies that looks wrong but is actually more right than the ones that look right. Every bug book will tell you that PMDs are yellow—possibly with a slight olivish cast—but if you start examining the Pale Morning Duns on the Frying Pan River, you'll notice that some of them are, in fact, a sort of pinkish color, although most are still the Ginger Quill yellow we're used to.

It's just an odd color variation that I was always too busy fishing to notice, but the trout really do seem to prefer it—or at least a fair number of those that will eat at all will bite the pink fly after refusing the yellow one. No telling why. Maybe the pink ones taste better.

I once thought this was just one of those cranky local patterns exclusive to the Frying Pan, but it has ended up working for me on a lot of different rivers that have Pale Morning Dun hatches. It could be that the pink phase is common in lots of places, but I've never been one of those who could spend an hour netting and examining mayflies while trout were rising, so I can't say for sure. I do know that several eastern fishermen have told me that when they can't make a regular yellow PMD work, they'll try a Light Hendrickson, which has a pinkish dubbed body. George M. LaBranche may have been onto

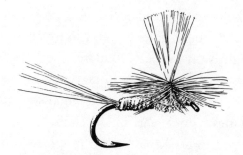

Frying Pan Pale Morning Dun Parachute

the same thing when he came up with the Pink Lady back in the 1920s.

I like the Frying Pan PMD in the collar-hackled, hen-winged version, and the parachute with a white turkey T-base wing and a dubbed thorax of pink rabbit fur, both in Sizes 16, 18, and sometimes 20 on the 94840 hook. I usually have some in my box that were tied by A.K. himself, because the gold-dyed hackle and tail he uses make it such a pretty fly, but when I tie my own, I use quills dyed as close to A.K.'s specs as possible, pale dun hen-hackle wings and the yellowest nat-ural-ginger tail and hackle I can find.

Biot Pale Morning Dun

Another Pale Morning Dun dry fly that works well for me is a standard parachute with a white turkey wing post, ginger tail and hackle, a body of yellow-dyed goose biot, and a dubbed thorax of bright-yellow rabbit fur.

I don't remember when I started dubbing the thoraxes on my parachute flies, but I remember why. When I first got into parachute flies—which I now believe in almost unreason-ably—they looked unfinished to me, so I started covering the thread base below and ahead of the wing post with fine figure-eight wraps of dubbing. Then, later, I started moving the wing post back slightly on the body to more of a thorax position.

I probably got both ideas from tying Vince Marinaro's Thorax Duns. The Marinaro patterns had split tails, divided hackle-tip wings set near the middle of the hook shank, and a hackle wrapped in an X pattern so it splayed out front and back like real mayfly legs. They were great flies, but they were goofy looking and hard to tie well, and a parachute hackle now does the same thing more easily.

I think larger parachute patterns on long-shanked 94831 Mustad hooks look especially good with the wing post set farther back and the thorax dubbed, but I also like it on smaller flies tied on standard-length hooks. This puts all the anatomical features more where they belong and allows the trim abdomen to taper into a fatter thorax, which also seems more realistic. Anyway, it's either a little stylistic quirk or I never learned how to tie a parachute fly the right way. Take your pick.

Pale Morning Dun No-Hackle

I seem to have about the same kind of luck with a PMD No-Hackle as I do with the Blue-Winged Olive. I'll try one when the current is slow and smooth enough to float it and

Pale Morning Dun No-Hackle

Quill-Bodied No-Hackle Pale Morning Dun

when nothing more conventional will work. It often makes a difference.

Saying it like that makes me think I should start with a No-Hackle more often when the conditions are right because it's a pattern that really does pull it out for me pretty regularly when things get sticky. Plus, a dry fly without hackle these days is dirt cheap to tie.

I make these just like I do the Blue-Winged Olives, except for color. The split tails are natural-ginger spade hackle, the bodies are either dubbed yellow rabbit or yellow-dyed goose biot or dyed quill with a dubbed thorax, and the wings are light goose shoulder.

(Why I don't tie a No-Hackle with a pink quill body à la A.K.'s PMD, I can't say. For some reason I only just now thought of it. I'll have to try that next season. It could be the secret, killer fly on the Frying Pan.)

Pale Morning Dun Barr Emerger

I like the Barr Emerger for a PMD emerger pattern. It was invented by John Barr of Boulder, Colorado, whom I've known for longer than I can remember, so when I was fishing a spring creek in Montana one year and needed a Pale Morning Dun emerger because the damned trout wouldn't bite any of my

Pale Morning Dun Barr Emerger

dry flies or floating nymphs I bought some of John's down at the local fly shop.

I got those first flies out of loyalty to a friend, because at the time I thought they were pretty funny looking. By the end of the day I'd caught lots of trout on them and lost all but one fly, so the next morning I bought another, bigger handful. By then they looked a lot better.

John told me that he first tied this fly while fishing Nelson's Spring Creek in 1975. He noticed lots of apparently intact brown nymphs floating on the surface with the yellow mayflies just beginning to poke out the front, and those were the ones the trout were eating.

The fly has a tail of four or five soft brown hackle fibers clipped flat on the ends; a brown dubbed abdomen covering the rear two-thirds of the hook shank; a thorax of yellow dubbing on the forward one-third; and a wing case of yellowish olive hackle fibers that are pulled over the thorax, divided, swept back along the body, and clipped a little past the thorax as legs.

John ties his on a Tiemco 101 hook. I use a 200R, Sizes 16 through 20, because I like the looks of it. Not that it matters.

This is a straightforward, businesslike pattern that's effective and easy to tie, and John says it's one the fish don't seem to get used to. That can happen. Sometimes the hot new pattern will wear out its welcome in a few seasons on a heavily fished catch-and-release stream, but this one has been a standard for longer than most with no end in sight.

There's also a Blue-Winged Olive emerger version of the same pattern that a lot of fishermen swear by. It's the same fly except the front end is tied from dun-colored dubbing and hackle. I've never tried it, only because I already carry too many Blue-Winged Olive patterns, but John made me promise to tie some and fish them this year.

Pale Morning Dun Palm Emerger

I've also started tying a Pale Morning Dun version of Roy Palm's Blue-Winged Olive emerger that uses John Barr's color scheme. It has a tail of reddish brown soft hackle fibers (there are some soft feathers around the tail of a ringneck pheasant that work well), a body of reddish brown–dyed goose biot, a yellow rabbit-fur thorax, and a yellow or natural-ginger hen hackle. Like the Olive version, I'll fish it either squeezed wet or greased so it floats, and I'll sometimes use it as a dropper behind a dry fly so I have some idea where it is on the water.

Pale Morning Dun Palm Emerger

Some of the toughest Pale Morning Dun hatches I've fished were on the Henry's Fork in Idaho (where you had better have everything in both No. 18 and No. 22) and on some of the spring creeks around Livingston, Montana, where large trout in small, clear streams are pounded unmercifully almost year-round by some very skilled fishermen.

I trust my favorite patterns and have done well with them, but I also make it a habit to stop at a local fly shop to buy a few examples of the hot pattern for that week, whatever it might be. Often enough they don't work any better than what I already have in my box, but sometimes they do, and every now and then something like the Barr Emerger works so well I decide to adopt it.

And of course sometimes the answer lies somewhere else entirely. Once Mike Clark, Ed Engle, and I were fishing DePuy's Spring Creek during a full-blown, boiling Pale Morning Dun hatch, and we'd probably have come very close to getting skunked if Mike hadn't figured out that what the fish really wanted was a Size 16 black foam beetle.

I borrowed one from Mike, and then during a break I drove into one of the fly shops in town and bought a handful. It was the same shop I'd been in a time or two already that week to buy flies. The guy behind the counter said, "Back again?"

Imitations

Blue-Winged Olive Dun

Hook: Mustad 94840, Sizes 16 to 24.
Thread: Light olive 8/0 or finer.
Tail: A small bunch of medium blue-dun spade-hackle fibers, cocked slightly upward.

Body: Pale olive rabbit-fur dubbing, dressed thin.
Wings: A pair of medium-dun hen-hackle tips, tied upright and divided.
Hackle: Medium blue-dun collar hackle.

Blue-Winged Olive Parachute

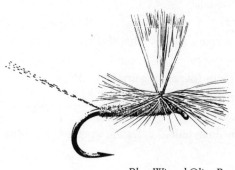

Blue-Winged Olive Parachute

Hook: Mustad 94840, Sizes 16 to 24.
Thread: Light olive 8/0 or finer.
Tail: A small bunch of medium blue-dun spade-hackle fibers, cocked
   slightly upward.
Body: Pale olive rabbit-fur dubbing, dressed thin.
Wing: Parachute post of white or medium-dun turkey T-base.
Thorax: Same dubbing as body.
Hackle: Four or five turns of medium blue-dun hackle wrapped parachute
   style.

Olive Dun Quill

Hook: Mustad 94840, Sizes 16 to 24.
Thread: Light olive 8/0 or finer.
Tail: A small bunch of medium blue-dun spade-hackle fibers, cocked
   slightly upward.
Body: One or two olive-dyed quills (depending on hook size), wrapped.

Wings: A pair of medium blue-dun hen-hackle tips, tied upright and divided.

Hackle: Medium blue-dun collar hackle.

Olive Dun Quill Parachute

Hook: Mustad 94840, Sizes 16 to 24.

Thread: Light olive 8/0 or finer.

Tail: A small bunch of medium blue-dun spade-hackle fibers, cocked slightly upward.

Body: One or two olive-dyed quills (depending on hook size), wrapped.

Wing: Parachute post of white or medium-dun turkey T-base.

Thorax: Pale olive rabbit-fur dubbing.

Hackle: Four or five turns of medium blue-dun hackle wrapped parachute style.

Blue-Winged Olive No-Hackle

Hook: Mustad 94840, Sizes 16 to 24.

Thread: Light olive 8/0 or finer.

Tail: Split tail of medium blue-dun spade-hackle fibers.

Body: Pale olive rabbit-fur dubbing.

Wings: A matched pair of light gray goose or duck shoulder feathers, tied upright and divided.

Thorax: Pale olive rabbit-fur dubbing.

Blue-Winged Olive No-Hackle—Quill Body

Hook: Mustad 94840, Sizes 16 to 24.

Thread: Pale olive 8/0 or finer.

Tail: Split tail of medium blue-dun spade-hackle fibers.

Body: One or two olive-dyed quills (depending on hook size), wrapped.

Wings: A matched pair of light gray goose or duck shoulder feathers, tied upright and divided.

Thorax: Pale olive rabbit-fur dubbing.

Blue-Winged Olive No-Hackle — Quill Body

Blue-Winged Olive Palm Emerger

Hook: Tiemco 200R, Sizes 16 to 20.
Thread: Pale olive 8/0 or finer.
Tail: A small bunch of wood duck flank feather fibers.
Body: One olive-dyed goose or turkey biot, wrapped.
Thorax: Pale olive rabbit-fur dubbing.
Hackle: Two or three turns of medium blue-dun hen hackle.

RS-2

Hook: Tiemco 200R, Sizes 16 to 20.
Thread: Pale olive or gray 8/0 or finer.
Tail: Two black moose body hairs, split.
Body: Pale olive or gray rabbit-fur dubbing.
Wing: A small piece of gray fluff from the base of a gamebird feather, clipped short.
Thorax: Dubbing to match the body.

Frying Pan Pale Morning Dun

Hook: Mustad 94840, Sizes 16 to 20 or 22.
Thread: Yellow 8/0 or finer.
Tail: A small bunch of golden-dyed or natural-ginger spade-hackle fibers, cocked slightly upward.

Body: One or two pink-dyed stripped quills, wound.

Wings: A pair of pale dun or cream hen-hackle tips, tied upright and
  divided.

Hackle: Golden-dyed or natural-ginger collar hackle.

Frying Pan Pale Morning Dun Parachute

Hook: Mustad 94840, Sizes 16 to 20 or 22.

Thread: Yellow or tan 8/0 or finer.

Tail: A small bunch of golden-dyed or natural-ginger spade-hackle fibers,
  cocked slightly upward.

Body: Two pink-dyed stripped quills, wound.

Wing: Parachute post of light dun or white turkey T-base.

Thorax: Pink rabbit-fur dubbing.

Hackle: Four or five turns of golden-dyed or natural-ginger hackle.

Pale Morning Dun Biot Parachute

Hook: Mustad 94840, Sizes 16 to 20 or 22.

Thread: Yellow 8/0 or finer.

Tail: A small bunch of ginger spade-hackle fibers, cocked slightly
  upward.

Body: One yellow- or golden yellow–dyed goose biot, wound.

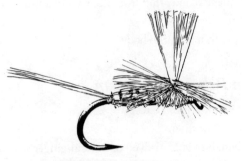

Pale Morning Dun Biot Parachute

Wing: Parachute post of light dun or white turkey T-base.
Thorax: Yellow rabbit-fur dubbing.
Hackle: Four or five turns of ginger hackle.

Pale Morning Dun No-Hackle

Hook: Mustad 94840, Sizes 16 to 20 or 22.
Thread: Yellow 8/0 or finer.
Tail: Split tail of ginger spade-hackle fibers.
Body: Yellow rabbit-fur dubbing.
Wings: A matched pair of gray goose or duck shoulder feathers, tied upright and divided.
Thorax: Yellow rabbit-fur dubbing.

Pale Morning Dun No-Hackle—Quill Body

Hook: Mustad 94840, Sizes 16 to 20 or 22.
Thread: Yellow 8/0 or finer.
Tail: Split tail of ginger spade-hackle fibers.
Body: Two yellow-dyed stripped quills or one yellow-dyed goose or turkey biot, wrapped.
Wings: A matched pair of gray goose or duck shoulder feathers, tied upright and divided.
Thorax: Yellow-dyed rabbit-fur dubbing.

Pale Morning Dun Barr Emerger

Hook: Tiemco 200R, Sizes 16 to 20.
Thread: Yellow 8/0 or finer.
Tail: A few soft brown hackle fibers, clipped flat on the ends.
Body: Dark brown rabbit-fur dubbing, dressed thin, covering the rear two-thirds of the hook shank.
Thorax: Yellow rabbit-fur dubbing.
Wing case and legs: A small bunch of yellow or yellowish olive hackle fibers, tied in behind the thorax, pulled forward as a wing case, and

then divided, folded back along each side and clipped just past the thorax as legs.

Pale Morning Dun Palm Emerger

Hook: Tiemco 200R, Sizes 16 to 20.
Thread: Yellow 8/0 or finer.
Tail: A small bunch of brown soft hackle fibers.
Body: One brown-dyed goose or turkey biot, wrapped.
Thorax: Yellow rabbit-fur dubbing.
Hackle: A few turns of yellow-dyed soft hackle.

Chapter 3

Medium-Sized Mayflies

I SPENT THE BETTER part of ten years and went through dozens of different fly boxes trying to figure out how to organize my flies in a way that made sense, not scientifically or aesthetically, but in a way that would work in failing light when the river was trying to push me off my gravel bar and I was fumbling for the box that held what I thought would be the right fly.

I ended up dividing mayflies into three categories that fit into three separate fly boxes: little ones (Size 18 and smaller), big ones (Size 12 and larger), and middle-sized ones (Sizes 14 and 16). Sure, a few patterns slop over from one box to another, but the system still works because I think size is the single most important thing about a fly, followed by shape and color, in that order. If I'm going to get caught short, I'd rather have the wrong fly in the right size than the other way around.

Hare's-Ear Parachute

When my medium-sized mayfly box is fully stocked, the most common fly in it is the Hare's-Ear Parachute because I go through them by the dozen. I first saw this pattern in *Tying Flies with Jack Dennis and Friends*, published in 1993, and it just looked good to me. For one thing, it's a white-winged parachute—a style of dry fly I began to get fond of in my late forties when I suddenly couldn't spot an Adams on the water like I used to. It's also simple, nondescript, buggy, a little scruffy by virtue of the hare's-mask dubbing, and it just looks like it ought to catch trout. If you fly-fish long enough, you begin to think you can see that in a fly.

According to the book, the Hare's-Ear Parachute was invented by Ed Schroeder, who also came up with the Schroeder Parachute Caddis, the Parachute Hopper, and other popular flies. (Jack Dennis is one of the few fly-tying writers I know of who properly attributes patterns whenever possible.)

Schroeder's pattern is tied on a standard-length hook with a split elk mane or elk hock tail, a dubbed body of hare's mask with a brown-thread rib, a white hair wing, and a grizzly hackle, but as I tied and fished the fly, I ended up changing it

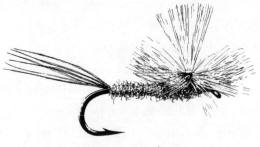

Hare's Ear Parachute

a little. I like it on the 2X long Mustad 94831, sizes 16 to 10, with a straight tail of bleached moose body hair, the thinnest dubbed body I can manage, either no rib at all or a fine brown one, and a wing post of white turkey T-base. Otherwise, it's the same fly.

I also tie a few of these on standard-length Mustad 94840s in Sizes 16, 18, and sometimes 20, and it's probably the most versatile dry fly I fish. At one time or another and in various sizes it has passed for all kinds of mayflies, from Blue-Winged Olives to Pale Morning Duns to Trico duns to Sulphurs to Red Quills, *Callibaetis*, *Flavilineas*, and Green Drakes, plus some dark midges, various caddis flies, and a spinner or two. It has also worked well as a search pattern on freestone streams when there was nothing obvious on the water and no trout were rising.

Ed Engle recently said he thought it might also suggest a hopper. (He pointed out that the way I tie it, on a long-shanked hook, it's a Schroeder Parachute Hopper minus the tent wing and legs.) I'd never thought of that, but I had to admit I'd caught a lot of trout on No. 14s and 12s fishing them as search patterns during the hopper season in late summer.

When I first started tying this pattern, I used black moose body hair for the tail, simply because I like that kind of tailing on larger flies. I discovered the bleached moose in a fly shop somewhere. It was a golden-brown color that would go perfectly with a hare's-ear-bodied fly, and it was a good bleaching job because the tips of the hair weren't singed or burned off. Later, Vince Zounek showed me how to get the same effect by putting black moose into a bath of drugstore hydrogen peroxide. This takes much longer than bleach—days longer—but it leaves the material in better shape, and the slow process lets you wait for precisely the color you're after.

I didn't notice any improvement in the performance of the fly when I went to the bleached moose tail, I just like the way it looks.

I've talked to people who say they have trouble making this fly float, but I think that's because they're tying the body too fat, which is easy enough to do with Hare's-ear dubbing, especially if you're used to tying your Hare's-Ear Nymphs on the beefy side. If I have trouble spinning the fur thin enough on the thread, I'll put a pinch of it back in the blender to make it finer.

Of course, this isn't the only magic, all-purpose dry fly, and lots of fishermen have something like it: the fly that they just have faith in. For many it's a Parachute Adams, for others it's something stranger, like a Lime Trude. This just happens to be mine, and I probably owe Mr. Schroeder something like a box of good cigars for inventing it.

Red Quill

The Red Quill is another fly that straddles the fly boxes. I tie most of mine in Sizes 14 and 16 on the Mustad 94831, but I also like to have some smaller ones on the 94840 in Sizes 16

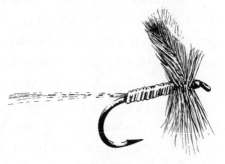

Red Quill Dun

and 18. A.K. recommends carrying Red Quills in Sizes 12 through 20, and that's probably not a bad idea. I have tied them as large as a Size 10, and they've worked as Brown Drakes and as mayflies known as Great Slate-Winged Drakes, which may or may not be a color phase of the Green Drakes, depending on which entomologist you consult.

Red Quills are a little like Blue-Winged Olives in that there seem to be a lot of mayflies around that fit the description: that is, a dark-colored mayfly with a brown or reddish cast to it like the Dark Hendricksons, Quill Gordons, March Browns, Mahogany Duns, and such.

The Red Quill most of us know was first tied by Art Flick. It's on a standard-length dry-fly hook with a medium-dun spade-hackle tail; a body of stripped, natural reddish brown quill; a divided wing of wood duck flank; and a medium-dun collar hackle.

The pattern I tie is usually on a standard-length Mustad 94840 hook because it looks more classic that way, and it's exactly like the Flick pattern except with the dun hen-hackle wings I copied from A.K.'s Red Quills. I've never cared for rolled flank wings because they're too sparse—once they're Ginked, they almost disappear—while the hen-hackle wings give a bigger silhouette and are more like the color of a real mayfly's wings.

I also like the Red Quill Parachute, which is the same pattern tied parachute style with a white turkey T-base wing and a thorax dubbed from reddish brown rabbit fur. In fact, I must like it better than the collar-hackled version because it's usually the one I try first when I want a Red Quill, possibly because it's easier to see. Still, I carry both patterns, and I use both.

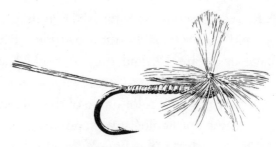

Red Quill Parachute

The Parachute Red Quill is a fly I could probably do without if I wanted to be efficient, because the Hare's Ear Parachute is so similar. But there are times when trout are more likely to tumble for the trim, segmented quill body, and, anyway, it's such a handsome fly, I just enjoy fishing it.

Sulphur

I don't know if the Sulphur mayfly we have here in the West is the same as the eastern fly or not. I once asked a bug guy about it and got an answer that was much longer winded than I'd hoped for. (I was sort of looking for a yes or a no.) I did come away with the impression that in various parts of North America there are various sized yellow mayflies, which I knew already.

I do know that if you ask some Pennsylvania limestone creek fishermen, they'll say the Sulphurs come in hook Sizes 12 or 14 down to 18 or 20, while here in the West most call the larger ones Sulphurs and the smaller ones Pale Morning Duns. Older fly casters are more likely to call them all Light Cahills or Ginger Quills and wonder what all the fuss is about.

Most of the Sulphur hatches I've seen in the West have been sparse, with few bugs on the water at any one time, and they're often mixed with other hatches, like Blue-Winged

Olives on the tailwaters or *Flavilineas*, Red Quills, caddis, or whatever on the freestone creeks.

But the trout must have a taste for them. One of the things we've learned on the Frying Pan River is that you can fish a Blue-Winged Olive hatch and pick up a few fish, but if you see even one or two larger Sulphurs mixed in with the smaller bugs, you can usually switch to that pattern and catch more fish. And the trout will often seem more eager, too: more likely to just bite the thing instead of making one of those suspicious, slow-motion refusal rises.

The same goes for the small freestone creeks I like to haunt in the late summer and early fall. There can be half a dozen different flies on and over the water, but if one of them happens to be a medium-sized yellow mayfly, then that's a good pattern to start with.

My best Sulphur pattern is a slightly larger version of the Pale Morning Dun Parachute tied on a 2X long Mustad 94831 hook. The tail is ginger spade hackle, the body is dyed yellow goose or turkey biot, the wing post is white turkey T-base, the thorax is dubbed with yellow rabbit fur, and the hackle is ginger.

I tie these on Size 14 and 16 hooks, and if I'm not in a hurry, I'll often tie them with a split tail. I'm not sure why ex-

Sulphur Parachute

cept that I tied a batch that way once and just liked the way they looked. I usually don't go to that kind of trouble with any-thing except a No-Hackle, where the split tails serve the func-tion of keeping the fly from tipping over. When the hackle does that on a parachute, I think split tails are just for looks.

I'd been fishing this fly in the West for quite a while, but when A.K., Mike Clark, and I went to Pennsylvania a few years ago to fish the Sulphur hatches, I was happy to see that it worked pretty well out there, too—except for one evening on Spring Creek, when Carl Roszkoski had to wade downstream and give me one of the neat Sulphur emergers he ties so I could catch a fish before dark. As I said, no fly *always* works.

I somehow managed not to lose that emerger, and I plan to make a few copies of it to try on the Frying Pan next year, although I'll probably substitute turkey and grouse for the poly yarn wing and Antron trailing husk. No offense, Carl.

I also tie a somewhat fancy version of this fly, essentially a parachute with divided hen hackle wings. I first saw this winging style in *Selective Trout* by Swisher and Richards. It was listed without much fanfare or description as just one more way to dress a Paradun (not to be confused with the Caucci & Nastasi hair-winged *Compara*-Dun). It looked neat but seemed impossible until I sat down at the vise and fooled with it for an afternoon.

You tie in a pair of hackle tips, concave sides out, as if you were making a pair of standard divided wings. Then, holding the wings upright in one hand, carefully wrap upward on the wings with the thread, using light tension at first and getting tighter as you build up the parachute post. (Yes, you have to let go of the wings on every turn and, yes, it's a little awkward at first.) I take quite a few wraps to make the post good and stiff, and then I take two wraps around the shank to anchor the thread.

The wings almost always twist in the direction of the thread wraps, so I just grab them firmly and twist them back. It also helps to cock them away from you slightly as you take your wraps.

If your feathers have a good natural curve to them, they'll sometimes divide nicely on their own. Otherwise, you'll have to divide them with a finger or even take a loose, figure-eight thread wrap through them.

I'll usually step-tie these. Once the wing post is done, I'll put a drop of thin lacquer on the thread wraps to cement the wings in place and stiffen the post a little more. Then I'll set it aside and tie another. By the time I've tied half a dozen, the lacquer is good and dry on the first one and I'll go back and do the thorax and the hackle.

Except for that wing post, this is a simple parachute dry fly: ginger spade-hackle tail, body of dyed yellow quill or biot, divided hackle wing parachute post in a cream or off-white color (and also a few in light dun), yellow-dyed rabbit-fur dubbed thorax, and a ginger parachute hackle.

Divided Wing Sulphur Parachute

I don't tie these in multiples of dozens and I don't usually start with one, but sometimes spooky trout really seem to like that winging style. It really does look realistic on the water, and I don't think it's something the fish see very often.

I've actually tied flies with that divided wing parachute post as small as a Size 20—just to prove a point—but I usually save it for medium-sized or large flies, where it's a little easier to do.

Flavilinea

Flavilinea mayflies always make me think of southeastern British Columbia, where I've been fishing for a decade now with my friend and guide Dave (Speed Bump) Brown. This is an important late-summer hatch all over the Kootenay region, and in the time I've been fishing up there I've seen the West Slope cutthroats get marginally more picky about the flies.

At first, you could fish just about any Size 14 or 16 dry fly and do well. One year A.K. and I ran out of Flavilineas, then Hare's Ear Parachutes, and finally ended up using Royal Wulffs and Humpies, sometimes as large as a Size 10. So far these cutts haven't entirely lost their native innocence, but I don't think you'd do as well today fishing a No. 10 Green Humpy through a *Flavilinea* hatch.

Flavilinea Parachute

Back before fly fishers started speaking Latin, the *Flavilinea* mayfly was known as the Little Western Green Drake or sometimes as one of the many Blue-Winged Olives. It was usually mentioned in the same breath with the larger, "real" Green Drake, and it came off as less spectacular and not as important.

But I think it's a better hatch in some ways. It's more common, more dependable, it lasts longer, and it's less famous than the Green Drakes, so it doesn't draw god-awful crowds of fishermen.

I tie my Flavs on No. 16 94831 hooks, and sometimes on the same-sized 94840s for a slightly smaller version. My normal fly has a sparse tail of black moose body hair, a thin body of A.K.'s Blue-Winged Olive dubbing mix, a fine brown floss rib, a white turkey T-base parachute post wing, a dubbed thorax, and a medium blue-dun hackle. (And in every batch I tie there will be two or three with dun wings, just because now and then that makes a difference.) If I'm fishing anywhere in the western United States in the summer, I feel naked without at least a dozen of these.

As a backup—something else to try on a difficult fish—I tie the exact same fly except with a divided, dun hackle-tip wing post and dun spade-hackle tail.

Divided Wing Flavilinea Parachute

A collar-hackled Olive Dun Quill on a standard-length Size 14 or 16 hook will also work as a Flav (especially in faster water), and any of these patterns can double as a *Callibaetis*.

Royal Wulff

My friend Mike Price is a master with the Royal Wulff. He ties them beautifully, and he's been known to catch fish on them, in one size or another, under all kinds of conditions, including little bitty technical hatches on catch-and-release rivers. He has admitted that he does this in part just to piss me off.

For years I tried not to like the Royal Wulff. I prefer flies that look something like real bugs, and although some writers have tied themselves in knots trying to prove that the Wulff actually passes as a sparkly caddis emerger or a flying ant, we all know that Lee Wulff himself was right when he said it was "strawberry shortcake."

But the Royal Wulff probably ties with the Adams as the most popular dry fly of the past fifty years, and you just shouldn't ignore something like that. Also, fishing a pattern that works even though it violates all your well-considered

Royal Wulff

beliefs about flies seems profound and has a way of putting things into perspective.

I fish most of my Royal Wulffs in Sizes 14 and 16 as attractors on freestone streams, but I also have some Size 12s for bigger water, plus some No. 18s and 20s. I wish I could describe when and where a Royal Wulff will catch fish, but that's pretty much unpredictable. I've had them work when there were no bugs on the water and no fish rising, but I've also had them work better than a more accurate fly during a hatch. It's a mystery.

Price has always tied his Royal Wulffs with a tail of black-tipped, orange, golden pheasant tippet as an allusion to the original quill-winged Royal Coachman he grew up fishing. (Mike is one of those rare holdouts who insist on thinking of a Royal Wulff as a Hair-Winged Royal Coachman.) Lately, he's been tying them with a little bunch of black moose body hair under a sparser golden pheasant tail to make them float better while still keeping the fly's traditional look.

Most seasons Mike gives me a handful of his ties and they immediately go into the box, but I also tie some of my own in what I take to be the accepted style. I use a black moose-hair tail, a body of peacock herl with the red floss joint in the middle, divided white hair wings, and a brown collar hackle.

Calf or goat body hair is a good winging material for any hair-winged fly because you can comb it out and stack it, making a neater, more compact wing, but I often use the crinkly off-white hair from the bottom of a snowshoe hare's hind foot.

Snowshoe hare hair is a fairly obscure material that was briefly made famous a few years ago by Francis Betters as the wing on his pattern called The Usual, and some other tiers have since picked it up. I probably wouldn't have paid any attention to it except that some friends and I do a fair amount of snowshoe hare hunting around here in the winter, so the feet were common and free. (Snowshoes make a good stew or gumbo, but if you want regular old fried rabbit, stick to cottontails.)

This hair can be a little tricky to work with at first, but it really is interesting stuff. It looks white on the fly, but it actually has a pale grayish or yellowish cast to it. The tips are crinkly, like calf tail, so it's not very stackable, but it also makes a fairly bushy-looking wing without a lot of material. It's also finer and straighter at the butts than it is at the tips, so you can comb out the short hairs and tie it in without getting a great big bump under the wings. That's especially helpful on smaller flies. It's also full of natural waterproofing oils, so it floats like a cork.

Interesting stuff, as I said, but I think any white hair that you can get and that's comfortable to work with is fine.

Parachute Royal Wulff

I also tie a parachute version of the Royal Wulff on the Mustad 94831 in Sizes 14 and 16, sometimes with a hair wing,

Parachute Royal Wulff

sometimes with a conventional parachute post of white turkey. There's no telling why I do this except that I like Wulffs and I like parachutes. I use them when the mood strikes, usually between hatches on freestone creeks, and they've done well.

For a while I thought I'd invented the Parachute Royal Wulff, even though I had to admit secretly to myself that dozens, if not hundreds, of other tiers had probably also come up with such an obvious idea.

Then I came across it in Jack Dennis's *Western Trout Fly Tying Manual*, published in 1974, which just happens to be the first fly-tying book I ever bought. Jack even tied it the way I do (or I do the way he did), with the peacock herl wrapped around and in front of the wing to make a peacock thorax.

I've looked at a lot of fly-tying books since, but I remember going through that manual page by page and photo by photo back then, so I have to assume that the pattern floated around in my head for twenty years before it drifted back into consciousness, where I temporarily mistook it for an original idea. That probably happens more often than any of us would like to think.

Imitations

Hare's-Ear Parachute

Hook: Mustad 94831, Sizes 16 to 10, or Mustad 94840, Sizes 16 to 20.
Thread: Brown or tan 8/0 or finer.
Tail: A small bunch of bleached moose body hairs, stacked and cocked
    slightly upward.

Body: Thinly dressed dark hare's-ear dubbing.
Rib: Brown floss (optional).
Wing: Parachute post of white turkey T-base.
Thorax: Dark hare's-ear dubbing.
Hackle: Four or five turns of natural grizzly hackle, parachute style.

Red Quill Dun

Hook: Mustad 94840, Sizes 12 to 20.
Thread: Brown 8/0 or finer.
Tail: A small bunch of medium blue-dun spade-hackle fibers, cocked
   slightly upward.
Body: One or two natural reddish brown stripped quills, wound.
Wings: A pair of medium blue-dun hen-hackle tips, tied upright and
   divided.
Hackle: Medium blue-dun collar hackle.

Red Quill Parachute

Hook: Mustad 94831, Sizes 12 to 16, or Mustad 94840, Sizes 16 and
   smaller.
Thread: Brown 8/0 or finer.
Tail: A small bunch of medium blue-dun spade-hackle fibers, cocked
   slightly upward.
Body: One or two natural reddish brown stripped quills, wrapped.
Wing: A parachute post of medium-dun or white turkey T-base.
Thorax: Reddish brown rabbit-fur dubbing.
Hackle: Four or five turns of medium blue-dun hackle, parachute style.

Sulphur Parachute

Hook: Mustad 94831, Sizes 14 and 16.
Thread: Yellow 8/0 or finer.
Tail: Ginger spade-hackle fibers, either split or straight.
Body: One yellow- or golden yellow–dyed goose or turkey biot, wrapped.
Wing: A parachute post of light dun or white turkey T-base.

Thorax: Yellow rabbit-fur dubbing.
Hackle: Four or five turns of ginger hackle, parachute style.

Divided-Wing Sulphur Parachute

Hook: Mustad 94831, Sizes 14 and 16.
Thread: Yellow 8/0 or finer.
Tail: Ginger spade-hackle fibers, straight or split.
Body: One yellow- or golden yellow–dyed goose or turkey biot, wrapped.
Wings: A pair of white or light dun hen-hackle tips, tied in upright and wrapped into a parachute post.
Thorax: Yellow rabbit-fur dubbing.
Hackle: Four or five turns of ginger hackle, parachute style.

Flavilinea Parachute

Hook: Mustad 94831 or 94840, Size 16.
Thread: Olive 8/0 or finer.
Tail: A small bunch of black moose body hair, cocked slightly upward.
Body: Pale olive rabbit-fur dubbing.
Rib: Brown floss.
Wing: Medium blue-dun or white turkey T-base parachute post.
Thorax: Pale olive rabbit-fur dubbing.
Hackle: Four or five turns of medium blue-dun hackle, parachute style.

Divided-Wing Flavilinea Parachute

Hook: Mustad 94831 or 94840, Size 16.
Thread: Olive 8/0 or finer.
Tail: A small bunch of blue-dun spade hackle, cocked slightly upward.
Body: Pale olive rabbit-fur dubbing.
Rib: Brown floss.
Wings: A pair of medium-dun or white hen-hackle tips, tied upright and wrapped into a parachute post.
Thorax: Pale olive rabbit-fur dubbing.
Hackle: Four or five turns of medium blue-dun hackle, parachute style.

Royal Wulff

Hook: Mustad 94840, Sizes 12 to 20.

Thread: Black 8/0 or finer.

Tail: A medium-sized bunch of black moose body hair, cocked slightly upward.

Body: Fine peacock herl with a red floss joint.

Wings: Upright, divided wings made of the hair from the foot of a snowshoe hare, or calf or goat.

Hackle: Generous dark brown (or furnace) collar hackle.

Parachute Royal Wulff

Hook: Mustad 94831, Sizes 14 and 16.

Thread: Black 8/0 or finer.

Tail: A medium-sized bunch of black moose body hair, cocked slightly upward.

Body: Fine peacock herl with a red floss joint.

Wing: Parachute post of snowshoe hare, calf, goat, or white turkey T-base.

Thorax: Peacock herl.

Hackle: Four or five turns of dark brown hackle, parachute style.

Chapter 4

Large Mayflies

I DON'T KNOW HOW the name Drake came to be used for some large mayflies, but it must somehow date back to the old days in England, where a drake could be either a male duck or any adult mayfly and where, eventually, the two meanings must have gotten tangled together in the same word. Or maybe some fisherman, exaggerating the way fishermen do, described a certain mayfly as being as big as a duck.

For some reason, most mayflies that are called drakes are tied on hook Sizes 12 and 10 or, now and then, 8, but the biggest mayflies, the *Hexagenias*, aren't called drakes, and in parts of Michigan they're called caddis. No telling why.

However it happened, it's part of the language of the sport now, so even if you'd never heard of a Western Gray Drake or a Slate Maroon Drake, you'd know from the name that it was a good-sized fly but not an enormous one.

The best thing about drake hatches is that they'll often get larger trout than usual up to the top to eat. And even if that doesn't happen—or if the big fish aren't all wall hangers—the big mayflies are just a joy to see. And I mean *see*. After weeks or months of squinting to spot a Size 20 emerger on the water, it's a relief to watch your No. 10 dry fly with three-quarter-inch wings clearly drifting along at a range of forty feet.

The worst part of this is that many of these hatches are famous for just those reasons, and they can draw awful crowds.

The first time I fished the Green Drake hatch on the Henry's Fork in Idaho was also the first time I saw a really huge mob of fishermen on a piece of water. I've seen bigger crowds on smaller rivers since then, but that first time was a genuine shock. There must have been a hundred cars, trucks, and campers parked across from Mike Lawson's fly shop in Last Chance, and anyone in town not wearing rubber pants stuck out like a sore thumb.

But the Henry's Fork in that stretch is as big and easy to wade as a flooded parking lot, so we found some open water to fish, the hatch came off, and we caught some trout; but I've been a little suspicious of drake hatches on famous rivers ever since.

Some of the best drake fishing I've had on well-known water has been after the hatch, when the bulk of the crowd has moved on to the next hotspot, whether it's upstream a few miles or in the next state. When nothing much is happening, or even during a sparse hatch of something else when the trout are looking up but not exactly in a frenzy, you can sometimes fish a drake pattern as a search fly and get a surprising number of strikes.

The effect fades over time (what effect doesn't?), but for a while—sometimes weeks—after the hatch has blown through, the fish seem to recognize the big mayflies and remember that they're supposed to eat them. A friendly local on a famous Colorado river once gave me this advice about fishing the Green Drake hatch: "Drive upriver until you start seeing guides with clients. Then turn around, drive a mile back downstream and fish there."

But then not all drake hatches are famous. I've seen good ones on small streams from Colorado to Alberta and British Columbia and some points in between, so I'll carry some of the flies anytime I'm fishing new water in the summer. Even sparse drake hatches in small, freestone mountain streams tend to get the trout pretty wound up.

There are some small, little-known headwater creeks in British Columbia that have skinny but predictable Green Drake hatches in late summer. Sometimes you can just fish the water with a drake and get strikes, while at other times you have to wait until you see a bug or two before the fish begin to look up.

Many times there aren't a lot of trout, and sometimes the biggest, deepest pools are crowded with pods of migrating bull trout, so the cutthroats will have moved out into water that doesn't look as good. But they're always there somewhere, a few of them are surprisingly large, and when even a few of the big bugs come off, they always rise.

I've been to these creeks only a few times, but I can picture stretches of them as if I were standing there right now, and once or twice I've bolted awake in the middle of the night thinking I'm there again. Maybe that's happened to you.

It's the big, wild trout, the big bugs, the smell, the romance of distance and foreign money, and the gnawing fear of grizzly bears that make those memories as permanent as tattoos.

Green Drake

The Green Drakes are probably the most well known, or at least the most publicized big mayflies, and in this case the fly-fishing entomologists all say that the eastern and western flies are definitely different bugs.

Most say the eastern Green Drake is an *Ephemera guttulata*. It's a bigger fly than the western drake, usually about a Size 10; it's paler in color, and sometimes the evening spinner falls are more important than the hatches. The spinners are large, too—a Size 10 or even an 8—and have white or pale cream-colored bodies. An eastern Green Drake spinner is called a Coffin Fly—another great but mysterious fly name.

I've never fished this hatch, but on a trip east not too long ago I bought two flies, a dun and a spinner, just because they were so big and impressive. Good fly patterns can amount to character sketches of the real bugs. Holding one in your hand isn't as good as seeing the real insect on a stream, but it's the kind of preview that makes you want to do that someday.

Western Green Drakes are roughly a size smaller—about a No. 12—and usually darker. People fish the duns and emergers during the day, but you don't hear much about the spinner falls because they happen in the dead of night. I've never talked to anyone who's fished the spinners, but I'm sure there are a few maniacs out there doing it, and not talking about it much, either.

It's not entirely clear to me what the western Green Drake is. Most call it an *Ephemerella grandis*, but, depending on who you read, it could also be a *glacialis* or a *doddsi*, and some divide *grandis* into three different subspecies that are scattered around the West from New Mexico to the West Coast to Alaska.

I guess that's why I don't read much entomology. I mean, it's interesting enough, but all it really means to a fly fisher or tier is that the Green Drakes on some western rivers may or may not be slightly different in color, and that may or may not make a difference.

I like two of the patterns A.K. ties specifically for the Frying Pan River, and I've had them work well in other places, too, including other tough, catch-and-release rivers and those little cutthroat streams in Canada.

Frying Pan Green Drake and Biot Drake

The Frying Pan Green Drake has a tail of bleached moose body hair (A.K. uses blond elk), a body of grayish olive dubbing with a brown floss rib, dun hen-hackle divided wings, and a mixed collar hackle of dyed olive grizzly and medium dun. The tail is tied a little shorter than usual and the wings are a little longer, to match the proportions of the real bugs.

A.K.'s Frying Pan Biot Green Drake is the same fly except with a body of dyed olive turkey biot. Sometimes when trout on catch-and-release waters refuse or ignore the dubbed

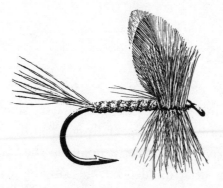

Frying Pan Green Drake

pattern, they go for the trimmer biot body. It happens often enough that I carry both.

Roy Palm also ties a Biot Drake with slate-colored, Wulff-style hair wings. It's a good pattern for faster water.

Of course, A.K., being a mad colorist and a perfectionist to boot, ties the dubbed body fly in three color phases and the Biot Drake in two. I tie them in one color, and when A.K. and I have fished Green Drake hatches together, I haven't noticed him catching a lot more fish than I do. Then again, that's not something I *would* notice.

A.K. also recommends trimming the collar hackles on these flies even with the hook point so the fly floats a little lower in the water. Sometimes I'll go that one better and clip the hackles even with the hook shank, but I do it only when I'm actually out fishing and need it that way. I don't trim the fly before then because I may want it fully hackled if I'm casting it to riffles or fast pocket water.

With the hackles untrimmed, the oversized wings on this fly will sometimes make it flop over on its side on the water, especially on windy days, but the real bugs do that, too—many of the big mayflies are sort of clumsy—and the fish like it.

Parachute Drake

I also tie both Drake variations in parachute versions with a divided dun hen-hackle wing post. I like parachutes in general, and I think they're especially good as drakes because the bigger the mayfly, the lower it floats on the water. And I am really sold on that divided-hackle winging style on a parachute, especially on bigger flies, where it's easier to tie.

Parachute Green Drake

For heavily fished hatches, where the trout can get pretty spooky, I like to have no less than four patterns to try: dubbed- and biot-bodied collar-hackled patterns, and the same two body styles in parachutes. I usually have more than that, though. Right now, in addition to the patterns I just described, my big mayfly box has some deer-hair extended-body Paradrakes, some Size 10 Green Drake Comparaduns, a couple of Rene Harrop's neat Hair Wing Drake duns, and some assorted hair-wing, Wulff-style flies, all store bought here and there, just in case.

And if none of those work, there's always the good-old Hare's Ear Parachute in a long-shanked No. 12 that I've made work as a drake on several western rivers. I fish Green Drake hatches the same way I do Blue-Winged Olives: Even if I can't get the fish to bite, I can stay busy changing flies for hours.

Harrop Green Drake Emerger

For some reason, I don't use a lot of Green Drake emergers, but one I do like is a Rene Harrop pattern I picked up in Mike Lawson's fly shop in Idaho years ago. It's called a Harrop Green Drake Emerger. It's a delicate, almost antique-looking, soft

Harrop Green Drake Emerger

hackle that's actually very similar to Roy Palm's Blue-Winged Olive emerger, although I couldn't say which fly came first. I tie it on a Tiemco 200R hook in a Size 10 or 12 with a wood duck flank tail, a body of dyed olive turkey biot, an olive-dubbed thorax, and a soft collar hackle of dyed olive grizzly hen. Harrop's original tie also had a few turns of soft black hackle in front of the dyed grizzly, but most of the commercially tied flies I've seen leave that off. I can go either way, but those few extra turns of black hackle really do dress up the pattern.

It looks way too simple and graceful to work, and almost all the other Green Drake emergers I've seen are fatter and have more body parts; but there's something about the simplicity of it that the trout seem to like. I usually fish these squeezed wet as a dropper behind a Green Drake dry fly.

Brown Drake

I suppose this is an odd fly to include in a list of favorite patterns because I hardly ever fish the hatch, but the few times I *have* fished it, it's been amazing and beautiful.

Those who know tell me that the Brown Drake I'm familiar with in the West is an *Ephemera simulans*, also sometimes called a March Brown. It's a big, brownish mayfly with a tan-colored underside and heavily mottled wings, usually tied on a Size 8 or 10 hook. The nymphs are big burrowers, so you find them in stretches of stream with silty or sandy bottoms, which aren't all that common in the Rocky Mountains.

Brown Drake hatches begin at dusk and last until after dark on quiet, slow-flowing stretches of river. Most of the ones I've seen started toward the end of the season's Green Drake hatches, and they were sporadic and short—from a few days to a week or two at most.

Fishermen are onto them, but Brown Drakes aren't as dependable as the daytime Green Drakes that are usually still on. The hatch hours are late and the fishing can be difficult, so they don't draw the same kinds of crowds.

Maybe the fishermen who chase Brown Drakes are a slightly different breed, or maybe a dark, silent river on a still evening just quiets us down. Whatever it is, you don't see a lot of restless bank stalking and compulsive casting. Mostly people lean on bridge railings or stand on the bank like herons watching the water. If the hatch doesn't come off, you'll probably see the same few guys standing around the next night.

If the hatch does come off, it's a desperate business. The entire thing may last only an hour, it's getting darker by the minute, and toward the end of it you'll hardly be able to see what you're doing. And this really is one of those hatches that bring the big, bruiser trout you never see in daylight up to the top to feed, so things can get pretty frantic. It doesn't help that sounds seem louder on a still, summer night, so even

Brown Drake Parachute

when you can't see anymore, you can still hear the trout out there gulping and splashing.

Spinners will be on the water at the same time as the duns after the first night or two of a Brown Drake hatch, but it's the big duns you notice first. It's half dark, but the current will be slow and smooth, and they'll look as big as ducks.

Everyone says the secret to this hatch is that on most nights the duns don't stay on the water for long, so the fish will be eating either the emerging nymphs or the spinners. Knowing that, you try to read the rise forms (are they dimpling the surface for spinners or boiling just underneath for the nymphs?), then pick either the nymph or the spinner and give it a good try over all the fish you can reach. This saves you from sinking into a paranoid whirlpool of fly changing. After all, it's a short hatch and you don't have much time.

What I've seen is some fish keying on one stage of the fly, some on another, and a few just eating everything that drifts by. I like to fish the duns because I can see them and because I just like a proper, hackled dry fly, so if I spot big sets of wings disappearing in a rise form, that usually turns out to be the fish I want. On the other hand, I'm not above tying a spinner or an emerging nymph off the back as a dropper.

The Brown Drake dun I tie is on a No. 10 Mustad 94831 hook. It has a tail of bleached moose body hair, a body of brownish tan dubbing blended from the cheeks of a hare's mask, a dark brown turkey-tail shell back, a brown floss rib, a divided-hackle wing parachute post made of mottled-gray Hungarian partridge flank, a dubbed thorax, and the same mixed brown and grizzly hackle you'll find on an Adams. It's probably a little fancier than it needs to be.

I've also tried a collar-hackled version of the same fly, but the fish seem to want this one floating flush, with its body right on the surface, and I think a parachute does that better on smooth currents than a collar-hackled fly, even if you clip the hackle level with the hook shank.

I like gray partridge for the wings because it copies the heavily mottled wings of the naturals so nicely. That may not make much difference in the dark, but even if it doesn't, the mottled wings make it a nice-looking fly, and the light partridge is a sort of ghostly, pale gray color that's reasonably easy to see in poor light.

The flank feathers are also short and stubby with strong, fat quills and a good natural curve to them, so they're perfect for that divided wing post.

I've seen Size 8 or 10 Red Quill Parachutes, March Browns, and even a big Adams pass as a Brown Drake dun, and I think a Hare's Ear Parachute in the same size would also work, although I've never tried it.

Brown Drake Emerger

Floating or slightly sunken nymphs really are important on Brown Drake hatches: more so than most because sometimes

Brown Drake Emerger

the trout will eat them exclusively and ignore the duns and spinners. In a pinch, I'll try almost any large unweighted or lightly weighted nymph, but I do tie a specific Brown Drake emerger. I use a Size 8 Tiemco 200R hook; a tail of mottled-brown partridge; a body of the same brownish tan dubbing I use on the dry fly; lashed gills of blue grouse or Hungarian partridge aftershaft feather; a shell back of brown turkey tail; a copper-wire rib; a dubbed thorax; folded over, or "drawn," brown partridge legs; and a wing case of the same brown turkey tail.

This is really just a standard shell-back sort of nymph except for the gills. Burrowing mayfly nymphs have prominent gills, and on the big Brown Drake nymphs they're like a row of flags down each side of the abdomen. The trout can't help but see them, especially when they're looking up at the emerging nymphs silhouetted against the evening sky.

I have wondered if the gills really make a difference, but it's an idle thought. As soon as I saw some of these big, burrowing mayfly nymphs, I wanted to copy the gills on my fly patterns, whether they make a difference or not. On the other hand, I have to admit that if I fished this hatch more often and went through more flies, I might have to try a simpler, more practical pattern.

Imitations

Frying Pan Green Drake

Hook: Mustad 94831, Size 12.

Thread: Olive 8/0 or finer.

Tail: Bleached moose body or blond elk.

Body: Medium to light olive rabbit-fur dubbing.

Rib: Brown floss.

Wings: Webby, gray hen-neck or hen-back hackle tips, tied upright and divided.

Hackle: Mixed olive-dyed grizzly and medium blue-dun collar hackle.

Frying Pan Biot Drake

Frying Pan Biot Drake

Hook: Mustad 94831, Size 12.

Thread: Olive 8/0 or finer.

Tail: Bleached moose body or blond elk.

Body: One olive-dyed turkey biot, wrapped.

Wings: Webby, gray hen-neck or hen-back hackle tips, tied upright and divided.

Hackle: Mixed olive-dyed grizzly and medium blue-dun collar hackle.

Parachute Green Drake

Hook: Mustad 94831, Size 12.
Thread: Olive 8/0 or finer.
Tail: Bleached moose body or blond elk.
Body: Medium to light olive rabbit fur dubbing.
Rib: Brown floss.
Wings: A pair of webby hen-neck, hen-back, or blue grouse flank feathers, tied upright and wrapped into a parachute post.
Thorax: Medium to light olive rabbit-fur dubbing.
Hackle: Mixed olive-dyed grizzly and medium blue-dun, parachute style.

Biot Green Drake Parachute

Hook: Mustad 94831, Size 12.
Thread: Olive 8/0 or finer.
Tail: Bleached moose body or blond elk.
Body: One light to medium olive-dyed turkey biot, wrapped.
Wings: A pair of webby hen-neck, hen-back, or blue grouse flank feathers, tied upright and wrapped into a parachute post.
Thorax: Light to medium olive rabbit-fur dubbing.
Hackle: Mixed olive-dyed grizzly and medium blue-dun, parachute style.

Biot Green Drake Parachute

Harrop Green Drake Emerger

Hook: Tiemco 200R, Size 10 or 12.
Tail: Wood duck flank feathers.
Body: Medium olive-dyed turkey biot, wrapped.
Thorax: Medium-olive rabbit-fur dubbing.
Hackle: Two or three turns of olive-dyed grizzly hen hackle with two
    turns of black hen hackle ahead of that.

Brown Drake Parachute

Hook: Mustad 94831, Size 10.
Thread: Tan or brown 8/0.
Tail: A medium-sized bunch of bleached moose body hair, cocked
    slightly upward.
Body: Brownish tan dubbing blended from the cheeks of a hare's mask.
Shell back: A strip of brown turkey tailfeather fibers.
Rib: Brown floss.
Wings: A pair of mottled-gray Hungarian partridge flank feathers, tied
    upright and wrapped into a parachute post.
Thorax: Brownish tan hare's-ear dubbing.
Hackle: Mixed brown and natural grizzly, parachute style.

Brown Drake Emerger

Hook: Tiemco 200R, Size 8.
Thread: Brown or tan 8/0.
Tail: A bunch of brown Hungarian partridge flank feathers.
Body: Brownish tan hare's-ear dubbing.
Gills: Lashed gills of partridge or grouse aftershaft feather.
Shell back: A strip of brown turkey tailfeather fibers.
Rib: Copper wire.
Thorax: Brownish tan hare's-ear dubbing.
Legs: Drawn legs of brown Hungarian partridge flank.
Wing case: A strip of brown turkey tailfeather fibers.

Chapter 5

Mayfly Spinners

ONCE, AFTER GETTING horribly skunked during a spinner fall, I developed what I thought was a perfectly logical theory about spent mayfly patterns. (I blame it on the two-and-a-half-hour drive home, alone, in the dark. If I'd been on the creek ten minutes up the road from the house, I probably could have avoided all this.)

Anyway, I got to thinking that my spinners were too perfect. I don't mean that I was too good a fly tier, I mean that spinner patterns in general were too neat and symmetrical. The first thing I did that night, when the trout wouldn't eat the fly I thought they should have, was look at the spinners on the water again.

Some of them—lots of them, actually—looked the way you'd expect a delicate little fly to look after it had been churned through a riffle that a grown man would have trouble standing up in. Their bodies were bent, their tails were cocked at awkward angles, their wings were mashed and folded. They were dead and they looked it. I'd seen bugs smashed on the grilles of cars that were in better shape.

The next day, I tied some free-form ugly spinners with crooked tails and wings in as many lopsided poses as I could come up with: one wing out and one swept back, both wings on the same side, that kind of thing. I even bent a few hooks sideways so the bodies would look sprung.

I thought they were beautiful, and I also thought I'd finally come up with something original: the killer dead-bug spinner pattern. (Later, I found out that a tier named Bill Thompson in New Hampshire was tying spinners on bent-shanked swimming-nymph hooks to get a similar effect.)

The next time I fished that same spinner fall, I tried one of these things and caught a couple of trout. Then, in the interest of science, I tried a conventional spinner and caught a few more.

That's pretty much how it's gone ever since. These things work about as well as the standard patterns, which can be anywhere from like a charm to occasionally not at all. Still, when I'm tying a batch of spinners, I now sometimes make a few ruptured versions. I'm convinced that some day I'll catch a big trout that wouldn't eat a normal spinner on one—and it will take only that one trout to prove the theory.

I fish spinner patterns like everyone else does: during spinner falls, when you have to copy at least the size and spent-wing posture of the naturals, if not the color. But on the advice of some old fishermen I've met, I'll also try them on solitary daytime risers, on the theory that they might be picking up leftover spinners from the night before. That's worked so well a few times that I'll now check eddies and backwaters along the banks to see if there are some old, dead spinners on the water. I've also now and then had spinners work during hatches of mayfly duns and midges. I think the fish must pick them out as cripples.

I once even caught a great big rainbow on a spinner fished like a nymph: ahead of a split shot in three or four feet of water. This was on Big Fishing Creek in Pennsylvania. The fish was near the bottom in a slow riffle, clearly feeding, and I'd spent more than an hour trying every nymph and wet fly I could think of, when Carl Roszkoski wandered by and asked me how I was doing.

I told him I'd just about decided it was useless, and he said that had been known to be true on that stream. Then we got to talking about last night's fall of Sulphur Spinners and about spinners in general. Carl said he thought spent spinners eventually got roiled under in fast water and that the fish probably fed on them underwater well after the fall was over.

It was the only thing I hadn't tried (because I'd never thought of it before), so I tied one of the Sulphur Spinners that had worked the night before onto my weighted nymph rig, and the big rainbow that had ignored all my nymphs for over an hour took it and I landed him, just like that. In the snapshot Carl took, the fish doesn't look quite as big as I remembered, but he's still pushing twenty inches.

You'll usually see spinner flights over riffles in the late afternoons and evenings, and sometimes in the early morning. If there's still a shaft of light down along the river, you might be able to see their wings glistening. It looks a little like a sun shower until you notice that the raindrops are falling and then bouncing back up again. If the light is off the water, you might spot their bodies dipping and rising against the sky.

Swallows darting over the water are also a good sign, and later bats: something you always want to look for late in the day.

Most spinners do their mating flights and lay their eggs over riffles, and sometimes trout will work up into the fast water to feed on them. When that happens, I can sometimes do as well with a parachute dry fly in the right size as I do with an actual spinner pattern. A parachute floats better than a spinner in rough water; the body lies flat on the surface like the naturals do, the hackle could pass for wings and I can sometimes manage to see the parachute post.

But the best spinner fishing is more likely to be after the fall in the quieter water downstream, where the fish don't have to fight the fast current and they can take their time feeding. As often as not, this will happen right at dusk or after dark, but even if the light is fairly good, you probably won't be able to see the bugs on the water because they're floating right in the surface film with their wings lying flat on the surface.

You can usually tell when trout are feeding on spinners by their rise forms. Spinner rises in slow currents are busy, but unhurried, and sometimes it's hard to see that the fish are even breaking the surface. It can also be difficult to pick out the rises of bigger trout, although sometimes a real pig will noticeably move some water.

I've seen fish dart around a little bit for spinners, but usually there are so many bugs on the water that they'll just get in a feeding lane and not move more than a few inches side to side, although they'll sometimes drift up and down in the current. If lots of trout are rising, you'll be tempted to flock-shoot them, but it's better to pick one fish and try to put the fly right in front of him.

You can't see your fly on the water anymore than you can see the real insects, so sometimes I'll pop the surface a little with my fly so I know where it is. I try to do that gently—with just enough of a blip so I can see it—and I try for a longer lead so I don't scare the fish.

I used to worry about slapping the water even lightly with a spinner for fear I'd sink it, but since that business on Big Fishing Creek, the idea of a sunken spinner doesn't bother me anymore.

Red Quill Spinner

The spinner I tie and fish the most is the Red Quill. At least half of the mayflies I know of, whatever color they are as duns, turn into rusty-colored spinners in their final molt. Sure, there are subtle differences, but lots and lots of mayfly spinners, from Size 10 down to the size of mustard seeds, are thin, graceful, and a rusty, reddish brown.

Even if they're a different color, most spinner falls happen in the evening or maybe the early morning when the light is low, so fishing a dark fly that throws a sharp silhouette against the background of a pale sky isn't a bad idea. In fact, there's a school of night fishing that says any fly fished after dark should be coal black for the same reason. Like all schools of fly fishing, the people who believe in it manage to catch some trout.

The Red Quill Spinner I tie is simple and very much like every other Red Quill spinner you'll see. It's on a Mustad 94831 for the larger sizes and on a 94840 from No. 16 on down. It has a split tail of white or cream hackle fibers, a body of stripped natural reddish brown quill, white or cream

Quill-Bodied Spinner with Hen Hackle Tip Wings

hen-hackle wings tied spent, and a thorax dubbed with rusty-brown rabbit fur.

Since you're copying nearly transparent wings on a spinner pattern, it's okay and maybe even preferable to use hen hackles that aren't webby enough to make good dry-fly wings. It's nice how that works out because it gives you something useful to do with those less-than-grade-A hen necks.

I like split tails on spinners because they look neat and because most of the spinners I see on the water really do have their tails splayed apart. But if I'm in a hurry or if I'm tying very small flies, I'll sometimes use a single, sparse tail and I haven't noticed any difference.

On really small flies, Size 22 and down, I'll sometimes forget about the quill body entirely and just use the finest rusty brown thread I can find, like 14/0. The bare hook shank is already as thick as the body on spinners that size, and I don't feel that I need to add any more bulk. Remember that spinners always have thinner bodies than duns of the same species.

Hen wings are usually the easiest and best for spinner patterns, but on some larger flies I'll now and then use hackle-style wings. That's where you wrap a tight, slightly oversized dry-fly–quality rooster hackle at the wing position, then sepa-rate it out to the sides with figure-eight wraps of dubbing. This takes a little longer than hen wings, but it's twice as durable.

You'll see a lot of synthetics used for spinner wings, and this is one place where that stuff makes a lot of sense. The wings on most natural spinners are sparkly and almost com-pletely clear. Light-colored hen hackle greased with fly flotant comes close, but if you're worried about exact imita-tion, it's still a little too opaque.

I haven't had a lot of trouble with this, but sometimes, in spring creeks and tailwaters where the trout have seen it all, a spinner with clear-plastic Zing wings will make a difference. I'll tie in a folded strip of plastic on top of the hook shank and stand the two sides upright and right together. Then I'll take curved scissors or a sharp pair of nail clippers with curved blades and trim both wings to shape at once. Finally, I'll sepa-rate them, lock them in place with figure-eight wraps, and tie in the thorax.

There are probably other clear materials that will work, but be careful not to get something that's too stiff. Stiff plastic wings on spinners (not to mention those snazzy burned wings on some dun patterns) can helicopter the fly and twist the leader.

Sulphur, Pale Morning Dun, Olive, and Trike Spinners

To my eye, Sulphur and Pale Morning Dun spinners are iden-tical except for size. I tie them in Sizes 2X long 16, as well as a standard-length 16, 18, and 20. If I know I'm going to be

Biot-Bodied Spinner with Wound and Divided Hackle Wings

fishing a PMD hatch on a spring creek somewhere, I may also tie a few No. 22s.

I tie them with split white tails, at least in the larger sizes; bodies of dyed yellow goose or turkey biot, cream hen or divided hackle-style wings (and sometimes a few of plastic); and a thorax of yellow dyed rabbit fur and yellow thread.

I tie my Olive Spinners in Sizes 16 through 22 or 24 in the same styles except that I use light dun for the tails and wings. The bodies are dyed olive quill or olive goose biot (or thread on the real little ones) and the thorax is medium olive rabbit dubbing.

For a Trike Spinner, a fly I don't use very often but wouldn't want to be without, I use white wings and tail, a dyed black

Biot-Bodied Spinner with Trimmed Plastic Wings

quill or goose biot body, and black thorax in Sizes 18, 20, and 22.

It's entirely possible that I don't need all those colors, or that I could at least get by with a light one and a dark one— say, a Red Quill and a yellow spinner. But there's something in a fly tier that makes him want to make a stab at copying the real insects, and it could be that color really *does* make a difference, especially on early evening or daytime spinner falls when better light makes the colors more visible to the fish.

Speckled Spinner

According to one of my bug books, there are twenty-eight species of *Callibaetis* mayflies in North America. They look a lot like Blue-Winged Olives, although they're usually a little bigger, and they all have heavily speckled, spotted, or mottled wings; hence the name Speckled Duns.

You'll now and then find *Callibaetis* on slow-flowing, weedy streams and rivers, but they're mostly lake and pond flies. They have multiple broods, so you're never really surprised to see them on still water.

I fish a Blue-Winged Olive pattern for the duns, but I like to tie the spinners with mottled wings, partly because I like the looks of them and partly because I think maybe the fish can see that mottling when the wings are flat on the water.

On a lot of the lakes I fish, heavy hatches of anything except midges are sort of rare, but the Speckled Spinner falls can be thick and the trout can really get on them.

The trick to fishing spinners on still water (or duns, for that matter) is to pick a gulper: a trout feeding steadily enough that he'll travel in a more or less straight line for at

Speckled Spinner

least a couple of feet. Then you try to lead him with your fly and wait for him to come up on it.

This takes a quick, accurate cast and the ability to see and remember where your fly landed because, once again, it's a spinner and you probably won't be able to see it on the water. Of course, half the time the fish will turn before he gets to your fly or take a natural right next to it. This is the kind of situation where the whole lake can be boiling with rising trout and you'll cast yourself into a cold sweat for a few fish.

I tie a Speckled Spinner on a Mustad 94831 or 94840, Size 16, with split tails of light dun hackle fibers, a body of stripped natural blue-dun quill, wings of mottled-gray partridge or hackle wings of finely barred grizzly, and a pale olive–dubbed thorax with light olive or gray thread.

Brown Drake Spinner

This is one of those bugs that's so impressive and distinctive, and the combination hatch and spinner fall itself is such a special event, that you can really get carried away when you're tying flies for it. The spinner is big enough to be tied on a No. 8

Brown Drake Spinner

or long-shanked 10 hook and has darkly speckled wings, prominent tails, and a two-tone body: dark brown on top, tan underneath. The fly could be essential on a combined Brown Drake hatch and spinner fall, and it's also one you could catch a great big trout on, so it seems worth the trouble.

I tie it on a Mustad 94831 No. 10 hook with split tails of brown spade hackle, a tan dubbed body with a shell back of brown deer or elk hair and a brown floss rib, spent wings of gray partridge, and a tan dubbed thorax.

The deer-hair shell back may not be entirely necessary, but I think it helps to float the big hook.

Then again, I've seen simpler patterns work. Not long ago Bob Scammell took me to a stretch of a little stream in Alberta where he said we might see a Brown Drake hatch. (Bob is a columnist, the author of a couple of good fishing books, and enough of a local expert that he's been known to have to hide his car when he goes fishing.)

I'm not sure exactly where we were (Bob took an unusually circuitous route to the stream), and I wouldn't be at liberty to tell you anyway, but I know the bridge we finally parked at was *not* the one known around there as the Brown Drake Bridge.

It was a cool night, right at dusk with no wind. It had been raining off and on and there were low clouds in the sky, but a rising moon was showing through an open spot with enough light to let us see a few big spinners in the air. One lone fish was rising at the bend where the stream turned out of sight.

I tried a nymph on that fish, but he stopped rising after two casts, so I reeled in and rested him. A few minutes later, he started coming up again. Bob said, "He's on the spinners," and handed me a fly he said would work. It was pretty dark by then, but I could see that this was a plain, large, soft hackle that was an overall nondescript brown.

I honestly didn't think much of the pattern, but I've learned the hard way to take the advice of kindly locals, even when it seems a little weird. Anyway, I tied on the fly and the fish took it on the second or third cast. There was a lot of splashing and cursing, but in a few minutes I'd landed a beautiful twenty-two-inch brown trout on a dry fly.

Bob took a picture of me holding it, and it's become my favorite hero shot. The stream, the woods, and the sky are all black, with a small, white moon just showing in the upper right corner of the frame. The fish and I are lit brightly by the flash, along with some tall, green grass in the foreground. The trout is big and handsome, and I have a stunned look on my face, as if I'd been caught by a hidden security camera in the act of stealing the fish.

By then it was full dark and no more trout were rising, so we walked back to the car. Bob hadn't even strung up a rod.

I guess it would be a better story if I'd caught that trout on my own fly, but I don't feel too bad. When I was rigging up,

I'd shown Bob my Brown Drake patterns. He said, "I don't use fancy flies like that myself, but those could work."

Imitations

Red Quill Spinner

Hook: Mustad 94831, Sizes 12 to 16, or Mustad 94840, Sizes 16 to 24 or 26.
Thread: Brown or tan 8/0 or finer.
Tail: Split white or cream spade-hackle fibers.
Body: One or two stripped natural reddish brown quills, depending on hook size, or reddish brown thread on the smaller sizes.
Wings: White or cream hen-hackle tips, tied spent, or white or cream hackle wound and divided, or plastic trimmed to shape.
Thorax: Rusty-brown rabbit-fur dubbing.

Sulphur and Pale Morning Dun Spinners

Hook: Mustad 94831, Size 16, or Mustad 94840, Sizes 16 to 20 or 22.
Thread: Yellow 8/0 or finer.
Tail: Split white spade-hackle fibers.
Body: One or two yellow-dyed quills or one yellow-dyed goose or turkey biot, wound.
Wings: White or cream hen-hackle tips, tied spent, or wound hackle divided, or plastic trimmed to shape.
Thorax: Yellow rabbit-fur dubbing.

Olive Spinner

Hook: Mustad 94840, Sizes 16 to 24 or 26.
Thread: Olive 8/0 or finer.
Tail: Split light dun spade-hackle fibers.
Body: One or two olive-dyed stripped quills, or olive goose or turkey biot or thread on the smaller sizes.

Wings: Light blue-dun hen-hackle tips, tied spent, or light dun hackle wound and divided, or plastic trimmed to shape.

Trike Spinner

Hook: Mustad 94840, Sizes 18 to 22.
Thread: Black 8/0 or finer.
Tail: Split white spade-hackle fibers.
Body: Black-dyed stripped quill or black goose or turkey biot.
Wings: White hen-hackle tips tied spent, or white hackle wound and divided, or plastic trimmed to shape.
Thorax: Black rabbit fur dubbing.

Speckled Spinner

Hook: Mustad 94831 or 94840, Size 16.
Thread: Gray or light olive 8/0 or finer.
Tails: Split light blue-dun spade-hackle fibers.
Body: Two stripped natural blue-dun quills.
Wings: A pair of mottled-gray Hungarian partridge flank feathers, tied spent.
Thorax: Pale olive rabbit-fur dubbing.

Brown Drake Spinner

Hook: Mustad 94831, Size 10.
Thread: Brown or tan 8/0.
Tails: Split natural brown spade-hackle fibers.
Body: Tan hare's-ear dubbing.
Shell back: Brown-dyed deer or elk hair.
Rib: Brown floss.
Wings: Mottled-gray Hungarian partridge flank feathers, tied spent.
Thorax: Tan hare's-ear dubbing.

Chapter 6

Midges

I THINK THE ONLY real advance I've seen in fly fishing since I've been doing it is the rise of midge patterns, and the availability of the little hooks to tie them on and fine leaders to tie them *to*. Before this miniature stuff came along, there were some hatches that just could not be fished.

By now, most hook makers have a few models that go all the way down to Size 26 or 28, but as late as the mid 1960s about the smallest hook you could find in any fly shop or catalog was a No. 18. Writers like Al McClane and Ernest Schwiebert now and then talked about hooks as small as a Size 22, but few normal people had ever seen such a thing.

This was also before entomological correctness caught on, so a midge was just as likely to be called a punkie, a no-see-um or, if it hatched in winter, a snow fly. You can still run into that refreshingly nonscientific approach if you look hard enough. I once fished out of a lodge in the Northwest Territories where the guides still referred to all aquatic insects as fish flies. You had your big ones and your little ones, your yellow ones and your green ones. It's not a bad way for a fly tier to look at it.

For a while, through the '70s, midge fishing was being promoted as graduate-level fly fishing: so advanced and refined that it was virtually a separate sport that required dinky little, otherwise useless rods, pocket-watch-sized reels, and leaders as fine as spider web. This was not something just anyone could do.

But it didn't take long for most fishermen to figure out that midge fishing was just fishing with smaller flies. You'd get yourself a spool of 7X tippet, maybe a pair of 2-power reading glasses down at the drugstore so you could see to tie on the flies, and you were in business. You didn't even need a new rod unless you just wanted one.

The few real tricks to it were just slight refinements of what you already knew, like being able to fish without seeing the fly on the water and taking care not to break off fish with fine tippets.

Like most other tiers, I had to work myself into small flies. It was just a matter of scale, but although the difference in shank length between a Size 14 and a Size 24 dry-fly hook is only about three-eighths of an inch, it seemed like a hell of a long way to go.

I took it slowly, tying No. 18 flies, then 20s, then 22s, then 24s. It seemed as if my fingers were swelling. I tied a lot, read some, got advice from other tiers, and even figured a few things out for myself, although I couldn't tell you which was which now. The best lesson was that small flies should be tied sparsely to make them easier, neater, and more realistic. As Darrel Martin said in his book *Micropatterns*, "In the world of insects, nature dresses her broods in scanty apparel."

For a while I got into those miniature midge-tying tools that some suppliers sell, but they didn't help and usually made things more difficult because they were so small I couldn't get a hold of them. Then one day A.K. asked me, "Where is it written that you have to use little tiny tools to tie little tiny flies?" I probably could have found someplace where it *was* written, but he had a point. If you're building a small house, you don't go out and buy a smaller hammer. Now I tie all my flies with the same familiar, comfortable tools.

A.K.'s Quill Emerger

The midge dry fly I fish most often on streams and rivers is A.K.'s Quill Emerger. This is one of those brilliantly simple patterns that achieves surprising detail with only four materials, that looks like the real insect on the water, and that's easy to tie. (I think ease of tying is a good feature of any fly pattern, but it's something I especially look for with midges because I can lose a lot of flies in a day's fishing.)

In the right size and color, a Quill Emerger seems to work almost anywhere in almost any conditions, although this may

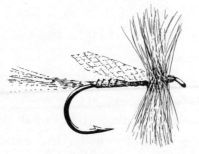

Quill Midge Emerger

be one of those self-fulfilling prophecies because I almost always try this fly first on a midge hatch.

It's an excellent hatching midge, but over the years I've discovered some other applications for it, too, usually by accident or out of desperation.

The olive-colored pattern in Size 18 or 20 has worked surprisingly well as a Blue-Winged Olive emerger, and sometimes better than flies tied specifically to imitate the actual mayfly. It's got the trailing nymphal husk, the cocked-back emerger-style wing, and it's the right color. The same pattern in ginger or cream has worked as a Pale Morning Dun emerger a few times, and I think that in the right sizes and colors it would pass for other mayfly emergers, although those two are the only ones I've tried.

I've also caught trout on various colors of Quill Emergers, fishing them as Olive, PMD, or Trike spinners. I don't really know why they work, although, as I said earlier, not all spinners are perfectly symmetrical on the water. (When I told A.K. about that he said the fish I'd caught didn't count because I was using the wrong fly. He may or may not have been kidding.)

The pattern has a trailing husk of wood duck or wood duck–dyed mallard flank, a dyed quill body, a wing made of slightly opaque plastic tied in as a single thin strip and trimmed to a point above or a little past the hook bend, and a sparse collar hackle.

A.K. cuts his midge wings from waffle-patterned plastic sheets used by graphite-rod makers, but any thin, slightly stiff plastic will do. I've also used white or pale dun wing feather segments, which work okay as far as the fish are concerned but aren't as durable or quite as easy to see on the water.

A.K. ties this pattern in olive—with an olive quill body and dun hackle—and with matching quill bodies and hackles in cream, dun, and black. I tie them in olive, ginger, and black, and I make mine as close to the way A.K. makes them as possible, except that I'll usually use a goose biot body instead of a quill. I think the biot is a little more durable, plus it lies a little flatter than quill on the hook shank.

For the most part I tie these in Sizes 18 through 22 on Mustad 94840 hooks, but I also like to have some smaller ones in Sizes 24 and 26. I'll also tie some No. 16s for lakes, where you'll sometimes find larger midge flies.

On the smaller ones I'm careful to use the narrowest biots, which you'll find at the base and at the tip of the goose feather. The wider biots are in the middle.

I don't tie the adult version of the same pattern because, as A.K. pointed out in his book *A.K.'s Fly Box*, if you decide you need the adult, all you have to do is clip off the trailing husk.

Lake Midge

I originally tied this pattern for a trout lake near home that had good midge hatches in March. This is a time of year when fishing weather and hatches are about as squirrelly as they get,

Lake Midge

but the lake was exactly a three-minute drive from where I lived at the time, so I could drive out there, glass the water from the cab of my pickup with the heater going, and if the fish weren't rising, I could turn around and go home. Lazy, but damned efficient. It got to where people were calling me to see if the hatch was on.

These were large, Size 16, olive-colored midges that would hatch and then skitter furiously across the water, buzzing their wings before taking off. They didn't go very fast, of course, and sometimes they'd go a few feet, stop, rest, and then do it again.

If the trout were dimpling and boiling lazily, I could occasionally get them on an emerger fished in the film, or on a slightly sunken pupa fished with the slowest possible retrieve, just enough to keep the line from going slack. But other times, often later in the hatch, the fish would get on the adults and eat only flies that were skittering slowly and deliberately across the surface. I could tell when they were doing that even before the truck stopped rolling because the rises were splashy and excited looking, the way trout act when they're feeding on a caddis hatch.

A couple of different midge patterns worked well enough, but they'd all sink too easily, regardless of how slowly and carefully I dragged them across the surface. After a couple of retrieves they'd be waterlogged and I'd have to tie on a fresh one.

So one night I tied some No. 16 midge flies with bodies of olive-dyed quill, half-spent dun hen wings, and a much bushier grizzly hackle than you'd expect to see on any midge pattern. I'd watched a few of these bugs motor past me as I stood thigh deep in the lake, and all I could see was a pale olive body with a gray blur at the front. I remember thinking briefly that all I'd have to do to get the same effect would be to clip the tail off a No. 16 Adams, but by then I'd already tied the flies, so it was too late.

The next day at the lake the flies floated better than anything I'd been using, and they caught some fish. Since then I've tied the pattern in sizes down to No. 22, and I use it whenever midges are skating around on the surface and the fish are chasing them. Once I was onto this, I realized it happens more often than I thought, on lakes from the foothills to timberline, and I'm told that the same flies on lakes in Ireland are called buzzers.

Chironomid Pupa

My friend Chris Schrantz showed me this midge pattern years ago. At the time, he was guiding at Elktrout Lodge near Kremmling, Colorado, and he said this was the killer midge pupa on the lakes and ponds around the valley where some large, smart trout were sometimes caught. Chris thinks the fly was originally tied by Marty Cecil, the head guide at Elktrout. I liked the looks of it and I do a lot of lake fishing, so I tied a few.

The first time I fished it was in a snowstorm in early April. This was on a trout lake just north of here that some friends and I like to fish in late March and early April, when the

Chironomid Pupa

trout first get on the midges in a big way and before too many people have fished there.

The best midge hatches naturally come off on the worst days. You want low, brooding clouds, temperatures in the high 30s, a cold breeze, and, ideally, a chilly drizzle or light snow. You dress as well as you can for it, but you're not hoping for comfort, you're just hoping not to die of exposure. However many trout you catch, you can't help thinking that this will be more fun to remember than it is to live through. As I said, we do it every year, sometimes four or five times.

The first day I fished Chris's pattern probably wasn't a good test because the weather was ugly, the hatch was heavy, and the fish were just stupid. I don't know how many trout I caught, but it was more than I'm used to. I'd tied half a dozen of those flies and was out of them by early afternoon.

I break off flies on these trips because I'm stiff and clumsy from the cold, or because my fingers are too numb to retie the fly to the leader after a few fish, or because the fly line has frozen in the guides. Whatever, I break off flies.

When I ran out, I tried something else and caught some more fish but not as many more. Then again, maybe the hatch was winding down.

I've done well on lakes with this fly ever since, though never quite as well as that first time, and just as I always try an A.K.'s Quill Emerger on rivers, I try this one on still waters.

I tie it on a Tiemco 200R, Sizes 16 to 22, or sometimes a Mustad 94840 in smaller sizes, although I think it's better looking on the 200R. That's also a slightly stronger hook for big trout.

The Chironomid Pupa has a sparse, short tuft of a tail made of poly yarn, a body of brown wild turkey-tail fibers (one to three, depending on the size of the hook), a counter-wrapped rib of fine copper wire, a short thorax of fine peacock herl, and a wing case and short post wing of more poly yarn.

I fish this one wet, either on a dead sink or with a very slow retrieve, and sometimes I'll use it as a dropper behind a lake midge.

This fly worked so well for me that I began tying a light version: the same pattern except with a body of pale yellow goose biot. I don't use a wire rib on this one because I think the biot is more durable than the turkey barbs and doesn't need reinforcing. I usually lose these things before they fall apart, anyway.

Sometimes the two patterns are more or less interchangeable, and sometimes the fish prefer one over the other (and of course sometimes they won't bite any fly in my box). So far I've resisted tying them in the full spectrum of possible midge colors.

Generic Midge Pupa

After looking at midge pupa patterns for decades, I've come to believe in the generic pupa: a sparse little body of almost anything you can think of with a short, sparse, little thorax of almost anything else. If there's a material that can go on a very small hook without building up too much bulk, some fly tier somewhere has used it to make a midge pupa.

The best of these are usually the simplest, but at the same time, the most minute details can make a huge difference. That seems unlikely and even unfair with such small flies, where, logically, you'd think you could get away with a light one and a dark one in a few sizes. But in my experience something like the presence or absence of a wire rib or a fuzzy body versus a smooth body on otherwise identical flies makes more difference than it should.

Often enough it's a matter of imitating the size and color of the naturals (the way some of us think it *should* be), but other times the weirdest things work. Red is a fairly common color for immature midges, but I've seen red floss and peacock flies work on days when you could seine the water and get nothing that wasn't a dirty cream or pale olive.

Many tiers just put materials together on a basic body-thorax model and try them. Most of those flies are anonymous and short lived, but some get to be personal favorites or secret weapons and a few become established patterns with names. If it has a body of copper wire and an ostrich or peacock herl thorax, it's a Brassie (or, in Colorado, a South Platte Brassie). A quill body and an Australian possum thorax make it an A.K.'s Quill Midge Larva. A white floss body with a copper wire rib is a Miracle Nymph, but the same fly with a rib of red thread is a Candy Cane Midge. With no rib at all, it's a String Thing.

(Some tiers say that, technically, a fly pattern with an abdomen and thorax is a pupa, and if it's just a wormlike body it's a larva. Others of us don't make the distinction, and I'm not sure the trout do either.)

I've tied dozens of different versions of the generic midge pupa that lived in the box for a few seasons until, nine times out of ten, they turned out to be either completely useless or at least no better than something I already had. I've added features such as a little bit of something at the tail for anal gills, thin beard hackles, shell backs, and so on. Nothing seems to have made much difference, with the possible exception of a little clipped tuft of fluff from the base of a feather (almost any feather) tied off the back. When I tie up a batch of midge pupae, I always tie a few with anal gills, just to give me something else to try.

Not long ago Ed Engle showed me how to tie white floss–bodied midges with different-colored thread underbodies to get subtle, translucent shades of pink, olive, brown, and such. I tried that, and the flies definitely look nice, but I haven't fished them much, so the jury is still out.

As it stands now, I pretty much depend on four pupa patterns tied in as many sizes as I can manage from 16 or 18 on down. In hooks sizes down to 22—which is as small as they make them—I like the Tiemco 200R because it gives the flies a realistic wormy, humped look. For flies smaller than that I use Mustad 94840s.

South Platte Brassie

According to an article by Ed Engle in *Fly Tyer* magazine, the South Platte Brassie was originated by Colorado tiers Gene Lynch, Ken Chandler, and Tug Davenport in the 1960s. The original pattern was tied with a body of copper wire and a short thorax made of black heat-shrink tubing.

At the time, you had to tie your Brassies fresh for every trip because the bare copper wire would quickly corrode and lose its

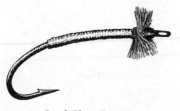

South Platte Brassie

flash. (When they were marketed in the '60s, they came in clear gelatin capsules to keep them bright.) The copper wire we use now has a thin, clear plastic coating, so it doesn't corrode and you can tie a five-year supply of Brassies if you want.

In the hands of most tiers, the chunky-looking heat-shrink tubing disappeared almost immediately and was replaced by a short thorax or collar of peacock or ostrich herl. Some people have added stubby wings and other features to the fly, and they're also now tied with red or green wire, but the original copper-bodied Brassie is still pretty much the standard.

I sometimes tie mine with fine peacock herl at the head, but I usually prefer two turns of gray ostrich.

Biot Pupa

I also tie a Biot Pupa that has a body of goose biot in pale yellow, olive, black, or natural Canada goose (whitish cream

Biot Pupa

with a distinct gray rib), and a thorax of dubbing that more or less matches the body color.

Hare's-Ear Pupa

This is a simple, pared-down version of the good old Hare's Ear Nymph: just a thin abdomen of hare's mask dubbing, a fine copper wire rib, and a sparse hare's mask thorax.

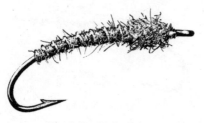

Hare's Ear Midge Pupa

I first saw fishermen using flies very much like this on Colorado tailwater streams in the 1970s. Many of the old hands said that a basic, no-nonsense Hare's Ear in size 18 or 20 would nicely split the difference between the various small mayfly nymphs and midge larvae and pupae that make up most of the trouts' diet on those tailwaters.

These were the guys who were catching a lot more fish than I was and who otherwise just seemed to know what they were doing, so I happily listened to their advice and followed it. (This was back when there were fewer fishermen and when the older guys were more likely to help out us young pups.)

I don't actually remember, but I may well have gotten my first Hare's Ear from a kindly, white-haired fly fisher—who would have been about the age I am now—somewhere on the South Platte River in the early 70s.

Miracle Nymph

Sometimes I'll also do well on a traditional gray Miracle Nymph with a black thread underbody, white floss overbody, and a copper wire rib. This would have to be called a midge larva, if you want to get technical. It's an effective fly, and I like the name, too.

Miracle Nymph

At this writing there are close to a dozen other pupa patterns in my midge box. Some I have yet to give a fair try; others haven't done much in several seasons and I'm getting ready to give them away. I like to get rid of flies that I haven't caught a fish on in recent memory. I do that either for neatness or to make room for future messiness, I'm not sure which.

Imitations

Quill Midge Emerger—Olive

Hook: Mustad 94840, Sizes 16 or 18 to 26.
Thread: Olive 8/0 or finer.
Tail: A small bunch of wood duck flank feather fibers, tied long.
Body: Olive-dyed stripped quill or olive-dyed goose biot.
Wing: A narrow, flat strip of waffle-patterned plastic tied in ahead of the body and clipped to a point above the hook bend.
Hackle: Medium blue-dun collar hackle.

Quill Midge Emerger—Ginger

Same as above except with ginger quill or yellow-dyed biot body, ginger hackle, and yellow thread.

Quill Midge Emerger—Black

Same as above except with black quill or black biot body, black hackle, and black thread.

Lake Midge

Hook: Mustad 94840, Sizes 16 to 22.
Thread: Olive 8/0 or smaller.
Body: One or two olive-dyed stripped quills, wound.
Wings: A pair of medium blue-dun hen-hackle tips, tied half-spent.
Hackle: A generous natural grizzly hackle collar.

Chironomid Pupa

Hook: Tiemco 200R, Sizes 16 to 22, or Mustad 94840, Sizes 22 and
   smaller.
Thread: Brown 8/0 or finer.
Tail: A sparse strip of off-white poly yarn, clipped short.
Body: One to three brown turkey-tailfeather fibers, depending on hook
   size, wrapped.
Rib: Counter-wrapped fine copper wire.
Thorax: Fine peacock herl.
Wing case and wing: A thin strip of off-white poly yarn tied in behind the
   thorax, pulled forward as a wing case, then stood up and clipped short
   as a wing.

South Platte Brassie

Hook: Tiemco 200R, Sizes 16 or 18 to 22, or Mustad 94840, Sizes 24 and
   smaller.
Thread: Black 8/0 or finer.
Body: Fine copper wire.
Head: Two turns of fine peacock herl or gray ostrich herl.

Biot Pupa

Hook: Tiemco 200R, Sizes 16 or 18 to 22, or Mustad 94840, Sizes 24 and
   smaller.
Thread: 8/0 or smaller to match body color.

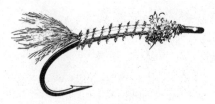

Biot Pupa with Tail

Tail (optional): A very small bunch of fluff from the base of a flank or hackle feather. Color to roughly match body.

Body: One goose biot in yellow, olive, black, or natural Canada goose, wrapped.

Thorax: Hare's ear or rough rabbit dubbing to more or less match body color.

Hare's-Ear Midge Pupa

Hook: Tiemco 200R, Sizes 16 or 18 to 22, or Mustad 94840, Sizes 24 and smaller.

Thread: Brown 8/0 or finer.

Body: Very thinly dubbed dark hare's-ear dubbing.

Rib: Fine copper wire.

Thorax: Dark hare's-ear dubbing.

Miracle Nymph

Hook: Tiemco 200R, Sizes 16 or 18 to 22, or Mustad 94840, Sizes 24 and smaller.

Thread: Black 8/0 or finer.

Body: One strand of white floss.

Rib: Fine copper wire.

Chapter 7

Caddis, Damsels, and Hoppers

I T MAY NOT BE VERY scientific, but I like to think of caddis flies as the opposite of mayflies: Mayflies are slow and float quietly on the water, but caddis flies are busy and blast off as if they actually understand that trout are trying to eat them. Mayflies are ephemeral—here and gone in a day or two—but caddis flies live for weeks, and some are always around. Mayflies in flight look like angels. Caddis flies look like moths on speed.

According to Rick Hafele and Scott Roederer in *Aquatic Insects and Their Imitations*, there are twelve hundred species of caddis flies in North America, more than all the species of mayflies and stoneflies combined. There are so many of them and they're all so similar that even fly-fishing entomologists don't try to use Latin names for the winged adults. We mostly describe them in terms of size and color and sometimes by when they hatch or how they act, as in October Caddis, Mother's Day

Caddis, or Traveling Caddis. Over the past twenty or thirty years we've dropped some of the more poetic, Old World–sounding names like American Sedge, White Miller, Autumn Phantom, Black Dancer, and such, which is too bad.

Hatching caddis flies leave the water quickly, and lots of fishing writers say the most efficient way to fish a caddis hatch is with a sunken emerger pattern. That's probably true, but a dry fly will also work, especially if you fish it with a little upstream twitch. (I first read about that in Leonard Wright's *Fishing the Dry Fly as a Living Insect*, and it's amazing how often it works.) On the other hand, when I've fished a caddis hatch with a dry and a dropper, I've often ended up getting two or three trout on the dropper for every one on the dry.

I always fish a dry fly to a mating swarm when the trout get on the egg-laying females. Most caddis flies lay their eggs on the surface, but I've read that a few swim underwater and lay them on the bottom. Some fishermen have been known to nip a split shot ahead of a dry caddis fly to sink it, and Gary LaFontaine has tied some patterns to imitate underwater egg-laying caddis adults. That all sounds like it would work, but I've never tried it.

I also fish a lot of caddis flies as search patterns. On most of the freestone streams I fish, caddis flies are around all summer and early fall. I think the fish get so used to seeing them and feeding on them that they'll eat one almost automatically if they get the chance.

St. Vrain Caddis

The basic body, down wing, and collar-hackle design has always been popular for caddis flies because it's effective, sim-

St. Vrain Caddis

ple, and easy to tie. It's used on Leonard Wright's Fluttering Caddis, Eric Leiser's Woodchuck Caddis, Jack Gartside's Pheasant Caddis, Wayne Buszek's King's River Caddis, and probably a dozen others.

One of the most versatile is A.K. Best's St. Vrain Caddis. A.K. started tying this years ago for a caddis hatch on Colorado's St. Vrain River, and it's been a standard on that drainage for decades now. But it also seems to work on any other stream, river, lake, or beaver pond where you see light-colored caddis flies.

This is basically an all-yellow caddis fly with a thin dubbed body of yellow-dyed rabbit fur, blond elk-hair down wing, and a ginger collar hackle, tied with yellow thread. This thing is brighter and yellower than most other caddis patterns and most natural caddis flies, too, but there's just something about it that trout buy into. It seems to me that a lot of successful fly patterns don't so much imitate an insect exactly as they exaggerate some feature of it: In this case it's the color. Anyway, I fish a St. Vrain Caddis to any light- to medium-colored caddis hatch or mating swarm, and also as an attractor.

Like A.K., I tie these in virtually all reasonable sizes from 12 or 14 down to 18 or 20 on a Mustad 94840. (Actually,

A.K. suggests using loop-eye or up-eye hooks for the smaller flies, which is probably not a bad idea.)

I like a fairly bright yellow dubbing right off a commercially dyed rabbit skin, and a hackle that's medium ginger. I'll use bleached elk hair for the wing, but I prefer the natural blond hair from an elk's rump. I still have the hide from the first elk I shot, and the rump patch is about the size of a dinner plate—maybe not quite a lifetime supply.

Bleached elk or deer hair really does work fine so long as it's the right yellowish blond color and hasn't had the tips fried off it during the bleaching process. If you're buying any patch of bleached hair in a fly shop, take it out of the plastic bag and flare it against the light from the window to make sure the tips are still there.

I also tie a dark version of the St. Vrain Caddis that's virtually the same as A.K.'s. The body is dubbed medium to dark olive, the wing is natural brown elk, and the hackle is mixed brown and grizzly. A.K. ties it with a plain brown hackle, but I picked up the mixed, Adams-style hackle somewhere along the line and I like the looks of it.

There's some nice chocolate brown hair to either side of the rump of a cow elk, but I prefer the slightly lighter hair up around the neck. It's not always possible, but when you're tying with hair, it's a luxury to have a whole skin to graze around on, looking for just the right color and texture.

Elk-Hair Caddis

Another dry caddis pattern I like is Al Troth's Elk-Hair Caddis. This has a dubbed body, palmer hackle, and an elk-hair down wing with a small head formed from the clipped hair butts. It's

Elk Hair Caddis

similar to the St. Vrain Caddis style, but it seems beefier and more substantial on the water, and there are days when that's what the trout seem to want. Sometimes I'll use the two styles interchangeably; other times trout really seem to prefer one over the other.

The Elk-Hair pattern also sometimes works for me as a windblown, egg-laying, or spent caddis when I clip the hackle flat underneath to make it float low in the water.

I tie what I believe is Troth's original pattern, with a hare's ear body, brown palmer hackle, and blond elk-hair wing. I also tie a light version in the same colors as the light St. Vrain Caddis: yellow, blond, and ginger. Both are on the Mustad 94840 in Sizes 12 or 14 down to 18.

Whenever I tie a fresh batch of any of these flies, I have to remind myself to use less hair than I'm inclined to use so the wing isn't too heavy, and to let off on the thread tension slightly at the base of the wing so as not to flare the hair too much.

I also try to tie a fairly trim dubbed body. Caddis flies seem chunkier than mayflies and it's easy to want to overdress them, as a lot of tiers do. It's just a matter of personal style, and any fly that comes out more or less the way you wanted it to is tied correctly, but I seem to have better luck with more sparsely tied caddis flies.

Labrador Caddis

After a few trips to far northeastern Canada, where we caught some enormous brook trout, A.K. and I took to calling any large caddis dry fly a Labrador Caddis. It's more of an inside joke than an actual pattern name, but this *is* my favorite big caddis fly, so I can't help thinking of it that way.

I first tied this about fifteen years ago for some prairie lakes in Wyoming that I used to fish with Jay Allman and a guy known as Wyoming Willie. These lakes had hatches of big—Size 8 or 10—traveling caddis that would hatch sporadically and then skitter off across the surface before they got airborne. They didn't flutter their wings like buzzer midges, they'd just run like hell across the surface.

There were never a lot of them on the water at any one time, but they'd peter off for hours, and when they were taxiing like that, they were easy, big bites for the trout. On days when the lakes were glassy, you could sometimes spot the bugs

Labrador Caddis

by their small wakes, now and then followed by the larger, faster wakes of trout. It could be funny or heartbreaking, depending on your mood.

What I needed was a large, buoyant, high-floating caddis fly with a brownish wing and a light body that could stand up to a lot of dragging across the surface. So I tied what is essentially yet another version of the old Henryville Special that Randall Kaufmann, writing in *Tying Dry Flies*, attributes to Hiram Brobst of Pennsylvania. This may not be the first caddis pattern with quill wings and both a palmer and collar hackle, but it's the oldest one I know of. By most accounts it's been around since the 1920s.

The fly I tied has a cream-colored dubbed body covered by a palmered ginger hackle, a tent-style down wing of mottled wild turkey with an underwing of dark bucktail, and a heavy ginger collar hackle. I originally used a Size 10 2X long 94831, but I've since tied them on 8s, 12s, 14s, and sometimes 16s.

For patterns like this I've settled on the tent-style wing tied from a single strip of wing quill that's folded over the body and then clipped to a V shape in back. The Henryville calls for two separate wing sections to be tied in on either side of the hook, but the tent style is quicker and easier for me to tie.

Later, I tied a dark version of the same fly on general principles. It has a medium-brown dubbed body and brown palmer hackle, brown turkey-tail wing, and brown collar hackle. These started out as lake flies, but I now also use them for big caddis hatches on streams and rivers. With the hackles clipped flat top and bottom, they'll also sometimes work as stoneflies.

Like half the other fishermen in North America, my first choice for a dry stonefly pattern is Randall Kaufmann's great Stimulator in the right size and color, but this pattern has been so tremendously popular for so long that it may be wearing out on some rivers.

Last summer A.K. and I were fishing on a small river in Alberta with Bob Scammell. There were some golden stones on the water, so I said to Bob, "What do you think, a big yellow Stimulator?"

"Try something else," Bob said. "These fish have been over-stimulated already."

I had a size 8, light Labrador Caddis in the box that already had the hackles clipped, so I tied that on and fooled a few fish. Of course, I'll never know whether the Stimulator would have worked as well or even better.

I've also done well at times fishing this big caddis as a midday search pattern, although the fish probably take it as a grasshopper.

Hare's-Ear Soft Hackle

A Hare's-Ear Soft Hackle is one of the great generic bug patterns. In one size or another I've used it as an emerging mayfly in rivers and as a *Callibaetis* emerger and damsel-fly nymph in

Hare's Ear Soft Hackle

lakes. It's also a good general search pattern, either by itself or as a dropper behind a dry fly.

I think of it as a caddis emerger probably because that's how I use it the most, and because it really is a fine, impressionistic copy of an emerging caddis pupa: an insect that, as Don Roberts once said, looks like a wet cat.

I've tried various ways of tying these over the years, but I finally settled on the easiest, simplest way. I use a Tiemco 200R hook and tie a short, carrot-shaped body of hare's-mask dubbing with either a copper wire or, on larger flies, a fine, flat, gold tinsel rib. I start the body above the hook barb, taper it and rib it, then tie in a slightly fatter thorax. Ahead of that is a sparse, oversized soft hackle of brown partridge on larger flies and Indian hen neck on smaller ones. When the fly is wet, the hackles should lie back past the hook bend.

They're unweighted in Sizes 18 and 20, but at Size 16 and above I put a few wraps of skinny lead wire under the thorax.

Back when tiers like Gary LaFontaine, Larry Solomon, and Eric Leiser first started to get more realistic with caddis pupa patterns, I got a little envious and tried to dress this thing up with short duck-quill wing pads underneath and long strips of bronze mallard trailing well past the hook bend as antennae. It looked good, but after a few seasons I convinced myself that neither of those things made a difference and went back to the sort of simple, traditional soft hackle that's been catching trout for the past five hundred years or so.

Bead-Chain Hare's-Ear Soft Hackle

But then a few years later I *did* come up with a variation of this that I still like and use. Caddis pupae can be pretty hefty, big-headed bugs, especially the larger ones, so I started tying

Bead Chain Hare's Ear Soft Hackle

the same basic soft hackle except with blued, bead-chain eyes with figure-eight wraps of hare's-mask dubbing through them for a head.

The silhouette is a little more like a real caddis pupa, and the bead-chain eyes give just enough weight to sink the fly an inch or two under the surface in current. I tie these in Sizes 16 through about 12 on the 200R in natural dark hare's ear and a muddy-olive rabbit-fur dubbing with the guard hairs blended in. I usually use brown partridge for hackles on both, but if I want to get real fancy, I'll use a hackle of olive-dyed gray partridge on the olive pattern.

I used to paint the eyes black on my bead-chain–eyed nymphs. I'd lash eyes to a few dozen hooks, paint them carefully with black enamel, then stick them in a Styrofoam coffee cup to dry overnight. The next day I could start tying.

It was pretty laborious, and the paint jobs weren't very durable. It didn't take long for the paint to chip and wear off, provided I didn't lose the fly in the first five minutes of casting.

I mentioned my paint problems to Mike Clark one day, and he asked, "Why don't you just blue them?" He then proceeded to show me how to do it. You start with silver-colored, nickel-plated bead chain, dip it in acetone to clean off the inevitable finger grease, and then dip it again into full-strength Jax Iron, Steel and Nickel Blackener, the stuff Mike uses to cold-blue the nickel-silver hardware on his bamboo rods. In a few seconds the whole chain turns a nice gunmetal blue black that won't wash off or chip.

Parachute Damsel

Fly tiers have been struggling with floating damselfly patterns for years, mostly without success, in my opinion. Some of the patterns are clunky monstrosities that might pass for dragonflies or small aircraft. Others are trim and delicate, like the naturals, with long, painstakingly scissored clear plastic wings. Some of these are beautiful, but they're also uncastable and they don't float.

Still, you have to try. Adult damsels aren't important very often, but they're a distinctive bug, and when they're swarming over a lake on a summer day and the trout are on them, nothing but a pretty fair copy will do. At least that's how it's always worked for me.

Parachute Damsel

I bought my first Parachute Damsels at a fly shop somewhere because I liked the simplicity of the original Bob Pelzl–Gary Borger pattern. It has a long, thin, deer-hair extended body, a half-moon–shaped parachute hackle, and a deer-hair wing case with a head formed from the clipped hair butts, like on the Elk-Hair Caddis. It uses only two natural materials, one of which—deer hair—is about as buoyant as anything you can tie with, short of cork.

I sat down to figure out how to tie the fly after I caught some fish the first time I tried one. The extended body goes on by the same method you'd use on something like a *Hexagenia* dry fly or a Swisher-Richards–style Compara drake. Tie in a bunch of deer hair on about the forward third of a standard-length dry-fly hook, lash it back along the hook shank with thread and continue with evenly-spaced wraps on out the bunch of hair to about one hook-shank length or a little more to make the extended body. Take a few tight wraps at the end and then spiral the thread back the other way to the beginning.

At this point I'll clip the deer hair flat on the end and put on a drop of lacquer.

Next, stand the deer hair butts up and wrap them into a parachute post. Tie in a slightly oversized hackle feather and dub the thorax. Then wrap a parachute-style hackle, fold the hackle backward, grab the deer hair, pull it forward, and tie it off for the wing case. (At first that final operation seems as if you need three hands, but you get the hang of it after a while.) Clip the butts of the deer hair to make a small head.

I tie these in blue and pale olive with light dun or cream hackles, and the only change I've made from the original is that I'll dub ahead of the parachute post with a dubbing material that more or less matches the deer hair, just to make the fly look a little more finished.

Any standard-length dry-fly hook will work for this pattern. I use a Mustad 94840 in about a Size 10 or 12. I also try to make the deer-hair extended body as thin as possible. This is a fly that's easy to overdress, and I always try to remind myself of what Craig Nova said about damselflies: that they're "like turquoise needles."

I saw this fly for sale almost everywhere for a few seasons, but it's since been pretty much replaced by similar patterns that have braided-plastic extended bodies and synthetic yarn for the post, wing case, and head. It's a little harder to tie with hair and probably not as durable, but I prefer what I take to be the original pattern, only because the main material comes from a real live deer.

Dave's Hopper

I have to admit that it took me a while to come around to the Dave's Hopper. The first one I saw was sometime back in the 1970s. A friend had bought it at a Trout Unlimited auction, and it was an original, tied by Dave Whitlock himself. I thought it was the prettiest trout fly I'd ever seen. I also

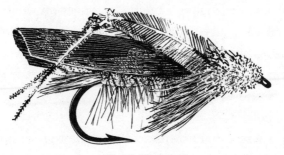

Dave's Hopper

thought it was a display pattern, not something anyone would ever tie to a leader and throw in the water.

But then, not long after that, I started seeing commercially tied Dave's Hoppers for sale at fly shops. I bought a few to try. When I ran out, I bought some more.

For a while I held them back as secret weapons—something to try when trout would flash other hopper patterns, but just didn't want to eat them—but eventually Dave's Hopper became my main grasshopper pattern, so I sat down and learned how to tie it.

It's a slightly complicated fly for me, with its red deer-hair tail, yarn body with a little yarn bump tied off the back as a short extended abdomen, clipped palmer hackle, bucktail or deer-hair underwing, turkey-quill wing, knotted legs, deer-hair collar, and clipped-hair head, but I can manage it well enough. The only change I ever made was in the legs. When the commercial flies switched from knotted pheasant-tail-fiber legs to knotted and clipped dyed-yellow grizzly-hackle stems, I did the same.

After some frustration, I found that the easiest way to clip the barbs short and even enough was to stretch the feather out on a piece of cardboard with my thumb and forefinger, and then trim the barbs with a razor blade. If you want to get fancy, you can shave the barbs down to almost nothing on the back part of the leg and leave them longer on what would be the thigh. Of course, then you have to make the overhand knot for the knee in exactly the right place.

If you tie the same pattern with a dark, reddish brown yarn body and everything else in black, you'll have a Whitlock Cricket.

I tie these on the Mustad 94831 or the slightly longer Tiemco 5263, in Sizes 14 through 6 or 8, but this is not a fly that goes quickly for me. I use a lot of them, especially in the smaller sizes, so I've been known to buy them, sometimes by the dozen. The commercial ties from Umpqua Feather Merchants are usually very good, and some other outfits sell virtually the same pattern under different names like Legged Hopper. I've ordered them through the mail a few times, but I'd rather go to a fly shop and carefully high-grade the best ones from the bins.

The only other fly patterns I regularly buy are the Umpqua Swimming Frogs I like for bass and pike, which, coincidentally, are also Dave Whitlock patterns.

Imitations

St. Vrain Caddis

Hook: Mustad 94840, Sizes 12 to 20.
Thread: Yellow 8/0 or finer.
Body: Yellow rabbit-fur dubbing.
Wing: Sparse blond or bleached elk hair, tied down-wing style.
Hackle: Medium-ginger collar hackle.

St. Vrain Caddis—Dark

Mustad 94840, Sizes 12 to 20.
Thread: Brown 8/0.
Body: Medium to dark olive rabbit-fur dubbing.
Wing: Sparse brown elk, tied down-wing style.
Hackle: Mixed natural grizzly and brown collar hackle.

Elk-Hair Caddis

Hook: Mustad 94840, Sizes 12 to 18.
Thread: Brown 8/0.

Body: Dark hare's-ear dubbing, thin.

Palmer hackle: Natural brown.

Wing: Sparse blond or bleached elk, tied down-wing style with the butts clipped to form a small head.

Elk-Hair Caddis—Light

Hook: Mustad 94840, Sizes 12 to 18.

Thread: Yellow 8/0.

Body: Yellow rabbit-fur dubbing, thin.

Palmer hackle: Medium ginger.

Wing: Sparse blond or bleached elk, tied down-wing style with the butts clipped to form a small head.

Labrador Caddis

Hook: Mustad 94831, Sizes 8 to 14 or 16.

Thread: Beige or yellow 8/0.

Body: Cream or beige hare's-ear or rabbit-fur dubbing, thin.

Palmer hackle: Medium ginger.

Underwing: A small bunch of brown bucktail extending to hook bend.

Wing: A strip of mottled wild turkey wing fibers, folded into a tent shape and clipped to a V in the rear.

Hackle: Medium-ginger collar hackle.

Labrador Caddis—Dark

Hook: Mustad 94831, Sizes 8 to 14 or 16.

Thread: Brown 8/0.

Body: Medium-brown rabbit-fur dubbing, thin.

Palmer hackle: Natural brown.

Underwing: A small bunch of brown bucktail extending to hook bend.

Wing: A strip of brown wild-turkey tailfeather fibers, folded into a tent shape and clipped to a V in the rear.

Hackle: Natural brown collar hackle.

Hare's-Ear Soft Hackle

Hook: Tiemco 200R, Sizes 12 to 18.
Thread: Brown 8/0.
Body: Dark hare's-ear dubbing, starting above the barb and tapering forward.
Rib: Fine copper wire or fine flat gold tinsel on larger sizes.
Thorax; Dark hare's-ear dubbing.
Hackle: Two or three turns of brown Hungarian partridge flank or brown to dark ginger hen hackle, swept back and extending just past the hook bend.

Bead-Chain Hare's-Ear Soft Hackle

Hook: Tiemco 200R, Sizes 12 to 16.
Thread: Brown 8/0.
Eyes: Small or medium blued bead chain, depending on hook size, tied in two or three hook-eye lengths back from the eye.
Body: Dark hare's-ear dubbing starting above the barb and tapering forward. Leave enough space to wrap the hackle between the end of the body and the eyes.
Rib: Copper wire or fine flat gold tinsel on larger sizes.
Hackle: Two or three turns of brown Hungarian partridge flank or brown to dark ginger hen hackle, swept back and extending just beyond the hook bend.
Head: Figure-eight wraps of dark hare's-ear dubbing around and through the eyes.

Bead-Chain Hare's-Ear Soft Hackle—Olive

Same as above, but with muddy-olive rabbit dubbing with the guard hairs blended in and a hackle of brown partridge flank, olive-dyed gray partridge flank, or hen hackle.

Parachute Damsel

Hook: Mustad 94840, Size 10 or 12.

Thread: Black 8/0.

Extended body: A thin bunch of blue-dyed deer body hair, tied in on the forward third of the hook shank and lashed back to the hook bend and on out one hook-shank length or more with cross-hatched wraps of thread. Clip flat at the end.

Parachute post: Deer-hair butts, stood upright and wrapped into a post.

Thorax: Blue dubbing.

Hackle: Five or six turns of light blue dun or cream hackle, parachute style.

Wing case and head: Fold the hackle back away from the hook eye, fold the deer hair forward, and tie off, leaving the clipped butts as a head.

Dave's Hopper

Hook: Mustad 94831 or Tiemco 5263, Sizes 6 to 14.

Thread: Yellow 6/0.

Tail: Red-dyed deer hair, short.

Body: Yellow yarn with a short loop over the tail as an extended abdomen.

Palmer hackle: Brown, clipped short.

Underwing: Brown bucktail or deer hair.

Wing: Mottled wild turkey-wing quill segment, rounded at the rear.

Legs: Yellow-dyed grizzly-hackle stem, knotted, with the barbs clipped short.

Collar: Natural deer hair.

Head: Natural deer hair, spun and trimmed to shape.

Chapter 8

Nymphs

I 'VE ALWAYS WANTED to be a dry-fly purist, but I've also always wanted to catch some fish, and I've never located the exact spot where the former leaves off and the latter begins. The only thing I'm really clear on is that I like casting dry flies a lot more than chucking lead. I enjoy nymph fishing when I'm catching fish, but when I'm not catching fish, it can wear thin in a hurry. I've done a lot of exploring, bird watching, and leisurely coffee drinking on days when the trout weren't rising and the nymph fishing was slow.

I'm using the term *nymph fishing* the way it's usually understood here in the Mountain West, to mean fishing nymphs with added weight on the leader to get them down on or very near the bottom. I learned how to do that years ago from my old friend Ed Engle on the South Platte River in Colorado, where some say the method was invented. (I've also heard of six or eight other rivers where it was invented, not that it matters.)

Nymphing with weight is an effective way to fish that can get you into trout at times when a dry fly never will, and that will often get you bigger trout, even during a hatch when the fish are rising like crazy. It's legal in most places (though not all), and it's now pretty widely accepted, although I remember a magazine editor rejecting a story I wrote about nymphing with weight twenty-some years ago because it was no better than bait fishing and no such crap was gonna appear in that publication on his watch. But, as I said, that was years ago.

It takes a lot of skill to nymph-fish well—you need concentration, great eyes, the ability to visualize in three dimensions (four if you count current speed), and mystical patience—but, on the other hand, it can be easy enough that a beginner with a little coaching and a big strike indicator can sometimes get into a few fish right off the bat. I think it's as popular as it is now because a lot of guides start their clients out with a brace of nymphs, a sinker, and a strike indicator, and a lot of fly fishers now assume that's what nymph fishing means.

I still do my share of dredging with weight on the leader—sometimes lots of weight, as much as it takes—but in the past few years I've tried to do it more sparingly. If there's anything wrong with this kind of nymph fishing, it's that it can be too effective. Lee Wulff once said that trout deserve the sanctuary of deep water, and I can't help thinking about that every time I nip three split shot onto my leader and dredge up a fish that might have started rising in an hour or two if I'd left him alone. Maybe there was a time when this didn't make much difference, but with the crowds you now see on popular rivers—not to mention the beat-up trout you sometimes catch—maybe the idea of letting the fish hide, rest, or feed undisturbed from time to time is worth thinking about.

So that's what I do—I think about it—and the more I do, the more I try to reserve dredging for those times when I really want a trout and can't get one any other way. It doesn't always work, but on the days when it does, I usually start thinking about the idea of sanctuary on about the third or fourth fish. Then I start wondering if I have the strength of character to quit and wait for a hatch.

But I do like to fish nymphs because they're so effective. The numbers are a little different in every book you read, but most of the people who study these things say trout do most of their feeding under the surface on nymphs, larvae, pupae, and other sunken bugs. Some of that goes on down on the streambed in the deep channels, but a lot happens in the top six inches of a trout stream, too, and during a hatch the fish have been known to get pretty single minded about it.

Sometimes they'll pick out almost nothing but the winged, floating flies, sometimes they'll seem to prefer the emergers; sometimes they'll stay just under the surface to grab off the rising nymphs; and sometimes they'll eat anything that comes along. And there are times when trout have individual preferences. You'll find a pod of them in the main current eating nothing but nymphs an inch or two under the surface, and one or two feet away, in an eddy, a few will be sipping nothing but floating duns. If you have good, fast eyes, you can tell what they're doing by the rise forms and save yourself a lot of fly changing.

Unweighted nymphs can be greased to float or squeezed wet to drift an inch or so under the surface, either alone or as droppers behind dry flies. (I like to tie my droppers long enough that if I miss a fish on the dry fly, I won't foul-hook him with the trailer. That usually works out to be about 18 or 20 inches.)

A lot of fly fishers who use droppers think of the dry fly as just a bobber or strike indicator for the trailing nymph. It works out that way often enough, but I still like to fish a dry fly that a trout might just eat. During a caddis hatch for example, I'll use something like a Hare's-Ear Soft Hackle behind the right-sized Elk-Hair Caddis.

If you weight nymphs with bead chain, the smallest dumbbell eyes, or wire wraps on the hook shank, you can get anywhere from a few more inches to another foot or so of depth in current and more in still water. I do a lot of that, and it works well. Most of the time, trout close to the surface are actively feeding or at least hunting for something to eat, even if they're shy about coming all the way to the top for a dry fly.

I didn't mention bead heads, but that's just because I don't like the looks of them. I know they're popular, I know they catch fish, and I've even used them a few times, usually to keep from insulting a guide or a friendly fisherman who insisted that a bead-head something-or-other was the only fly that would work. I just think they're ugly, so I can't bring myself to tie them, but it's nothing personal.

Some fishermen like to carry nymphs, floating nymphs, emergers, sub-emergers, cripples, still-born patterns, and trailing husk duns for every hatch, and that's definitely one way to do it. I carry specific emergers for a few familiar hatches, but I also get a lot of mileage out of plain old unweighted nymphs and standard dry flies.

If a floating or slightly sunken nymph doesn't get it, you can make a serviceable emerger from a standard nymph pattern by cutting the wing case at the back so it stands up like a stubby, half-emerged wing and maybe combing back some of the dubbing on the abdomen to form a trailing husk. You can also come at it the other way by clipping hackle and shortening the wings and tail on a dry fly. It's amazing what you'll try when the fish won't bite.

Most of the nymphs I use are easier and quicker to tie than most of the dry flies, and there's also a freedom to material selection that I like. Most good nymph material is a lot cheaper than good dry-fly hackle (though it's sometimes just as hard to find), and you can pick it by color, texture, and how it acts in the water without worrying about buoyancy. Apparently, the nature of reality is such that it's easier to tie something that looks like a bug and sinks than it is to tie something that looks like a bug and floats.

I don't tie nearly as many different nymph patterns as I do dry flies because I don't think I need to. Some aquatic nymphs have distinctive colors or shapes or some other obvious trait like prominent gills, but most are camouflaged in some shade of muddy brown or olive and are roughly carrot shaped with legs at the front. I get fancier than that with some patterns because I enjoy it, but sometimes I think I could do most of my nymph fishing—with and without lead—with a ragged Hare's Ear in the right size.

Hare's-Ear Nymph

My favorite nymph pattern by far is the Hare's Ear. I've seen these tied dozens of different ways, both as classic wet flies and as nymphs. They've all worked, and I have to think that has something to do with the material itself. Good hare's-ear dubbing is blended from the darker hair on a European hare's mask with the chopped-up guard hairs mixed in. There's black, shades of brown, tan, and gray, and textures from soft,

Hare's Ear Nymph

fluffy underfur to spiky guard hairs. It's a natural, neutral, nothing brown, like an old brush pile or a well-weathered road apple, and it's such perfect cryptic coloring that if you drop a Hare's-Ear Nymph on the ground right at your feet, you'll probably never find it. That is exactly the kind of coloration a nymph afraid of being eaten by a trout would want to come up with.

I usually decide how to tie my Hare's Ears by their size. On 18s, 20s, and 22s, I'll just tie a sparse tail of brown partridge or something similar, a tapered abdomen of hare's-ear dubbing with a copper wire rib, and a hare's-ear thorax with a dark wing case of brown wild-turkey tail fibers or charcoal gray Canada goose. If the dubbing on the thorax doesn't seem rough enough, I'll pick it out a little bit to suggest legs.

From Size 16 on up to Size 10 or 12, I tie the same fly with drawn feather legs—that is, with a hackle feather laid flat under the wing case so the barbs stick out on each side. I use

Hare's Ear Nymph (small)

brown Indian hen neck on smaller flies and brown partridge, ptarmigan, or some other gamebird flank feather on the larger ones. If I remember right, I first saw that style of nymph legs in Eric Leiser's book *Fly Tying Materials*, published in 1973. I don't know if he invented it or not, but I've always liked it. I think it's more realistic than a wound hackle, and it's no harder to tie.

I tie these on Tiemco 200R hooks with some lead wire under the thorax on all but the smallest flies. The unweighted versions go on Mustad 94840s so they'll be easier to float if that's how I want to fish them and so I can tell them from the weighted flies. Other tiers use different thread colors for weighted and unweighted nymphs so they can tell which is which by the colors of the heads.

Hare's-Ear Stonefly Nymph

I tie my biggest Hare's Ears—Sizes 8, 6, and sometimes 4—as stonefly nymphs. They have split tails of brown-dyed goose biot, a tapered hare's-ear dubbed abdomen with a brown turkey shell back, and a copper wire rib, hare's-ear thorax, drawn partridge legs, and a turkey-tail wing case.

Jack Sayers (a.k.a. Gray Ugly) told me about this fly years ago, and I jumped on it because it was yet another version of a

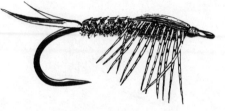

Hare's Ear Stonefly Nymph

Hare's Ear. Since then it has worked just noticeably better for me than dark stone nymphs tied in black or chocolate brown, although I do still carry a few black ones as backup.

Hare's-Ear Damsel

I also tie a lake and pond nymph that I think of as a Hare's-Ear Damsel. It's on a Daiichi 1870 Swimming Larva hook, Sizes 10, 12, and 14, with blued bead chain or the smallest-sized dumbbell eyes for weight. I especially like the hourglass–shaped eyes called Dazl-Eyes in the ³⁄₂- and ⅛-inch sizes.

This fly is not only a good dark damselfly nymph, but also an all-around still-water search pattern that slowly dives head first when you stop it and then rises when you pull on it. I think of it as a lake and pond fly, but I've also used it in streams as a large mayfly or stonefly nymph. The ones with bead-chain eyes are for shallow water or when fish are working close to the surface. The ones with dumbbell eyes are for deeper water.

It's tied with a brown partridge tail, thin hare's-ear abdomen with a fine copper wire rib, hare's-ear thorax with drawn partridge or Indian hen legs, a brown turkey wing case, a pair of eyeballs, and a small head formed by taking figure-eight wraps of dubbing through the eyes.

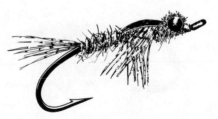

Hare's Ear Damsel (natural or olive)

I tie a green version of this with rough, muddy-olive rabbit dubbing, gray partridge tail and legs, and a dyed olive goose-feather wing case.

Golden Stone Nymph

Golden stones are common freestone stream insects around here, and for those I tie the same pattern as the Hare's-Ear Stone except in yellow. It has golden yellow–dyed goose biot tails, an abdomen of rough, golden yellow rabbit dubbing with a brown turkey shell back and copper wire rib, a yellow rabbit thorax, drawn legs of ptarmigan flank, and a brown turkey wing case.

At least that's how I tie them in the winter when I have plenty of time, or maybe even a little time, to kill. When I'm in a hurry, I'll tie the abdomen and thorax out of yellow furry foam. It's a little New Age, but the fish don't seem to mind and it cuts the tying time in half.

I'll usually weight my stonefly nymphs pretty heavily with lead wire wraps under the thorax, using wire roughly the same diameter as the hook shank. Sometimes I'll also wrap the forward half or two-thirds of the abdomen with a smaller-gauge wire to add a little extra weight and to form a more tapered underbody. Before I start tying the actual fly, I'll smooth the tapers behind both wire wraps with wraps of thread. It takes a little longer, but it makes a neater fly.

Golden Stone Nymph

Blue-Winged Olive Nymph

I tie two versions of the Blue-Winged Olive nymph, in keeping with my policy of carrying as many patterns as possible for that hatch.

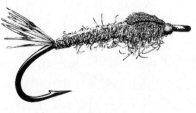

Blue-Winged Olive Nymph

One is about as basic as a mayfly nymph pattern can get. It has a short tail of either mottled or plain gray soft hackle fibers (blue grouse flank or dyed hen), a thin abdomen of pale olive rabbit-fur dubbing, a thorax of the same material tied a little fatter and roughed up a little for legs, and a wing case of gray Canada goose. I tie these from Size 18 down to 22 or 24, and I use the Mustad 94840 because more often than not I float them as emergers, so I want the light wire hook. I also tie some in Sizes 16 and 14 on a 200R as a *Callibaetis* nymph; on these I'll usually tie drawn legs of dyed dun hen-hackle with a few turns of fine lead wire under the thorax—just enough to keep the fly under the surface when it's drifted in slow currents or slowly retrieved in still water.

Tricolor Shell-Back Nymph

My other Blue-Winged Olive nymph is a close copy of A.K.'s Tricolor Shell-Back nymph. (I don't automatically buy all of A.K.'s ideas, but eventually I do tumble for most of them.) It's

Tricolor Shellback Nymph

the same as the Blue-Winged Olive nymph except that it has a shell back of gray goose with a dark olive or brown thread rib. A.K. ties his with a beard hackle, but I just use rough dubbing on the thorax for legs on smaller flies and replace the beard with drawn hackle on larger sizes.

The theory behind this fly is that most nymphs are darker on the top than on the bottom, which is true. Whether it makes a difference to the trout is a guess, but at least it's something else to try. The fly does work as often as not.

A.K. also ties what he calls a Sub-emerger version of the same fly with a stubby wing of gray yarn sticking out behind the wing case. I've never tried that one, but I have sliced the wing case at the back with a sharp pocket knife and picked out some of the thorax dubbing to get something like the same effect. This is one of those things I do only out of desperation, but it's been known to work.

Pheasant Tail Nymph

I think Frank Sawyer's Pheasant Tail Nymph is as brilliantly simple as the Hare's Ear and every bit as effective. The original pattern is tied from nothing but reddish brown pheasant tail barbs: a few fibers for the tail, a few more ribbed with fine gold wire for the abdomen and thorax, and a few more for the

Pheasant Tail Nymph

wing case. The original doesn't have legs (Sawyer said you don't need them because the real nymphs swim with their legs tucked in), but some tiers wrap the pheasant fiber wing case with the tips pointing forward and then fold some of them back for legs.

I've been tying my Pheasant Tails with a peacock herl thorax ever since I saw it on some commercially tied patterns and liked the looks of it. I also tie the wing cases from dark gray goose-wing fibers because the wing cases on real nymphs turn dark before they hatch, and I've replaced the gold wire rib with copper.

I tie these on the Tiemco 200R hook in Sizes 12 through 22 with a little weight under the thorax on the larger sizes. At one time or another I've had them work on all kinds of mayfly hatches—fished anywhere from right in the surface film to down on the bottom of the river—and the small ones have worked well during some midge hatches.

Sawyer also has another, less well-known pattern called the Gray Goose Nymph. It's tied exactly like the Pheasant Tail except with gray goose wing-feather barbs—not the biots but the long fibers on the opposite side of the feather. I haven't tried this one yet, but the next time I sit down to fill the nymph box I think I'll make a few so I can try them on the next Blue-Winged Olive hatch.

Peacock Nymph

For as far back as I can remember I've always carried some kind of peacock-bodied nymph, at first because just about all the good fishermen I knew had one, later because I'd come to depend on it in a haphazard sort of way. Trout just like the weird green metallic fuzziness of peacock, and that's why it's been cropping up on trout flies at least back to the old Coachman wet fly.

I usually fish Peacock Nymphs as search patterns in lakes and streams, and sometimes I'll tie one on as the dropper behind a Royal Wulff when no fish are rising, just to fill out the strawberry shortcake motif. I've had this work so well on brookies and cutthroats in small mountain streams that I started to feel guilty.

I almost never use a Peacock Nymph when I'm fishing to a specific hatch, but sometimes, after I've tried everything else that seemed reasonable, I've resorted to one in about the right size and had it work. I don't know the principle that's in operation here, but it's the same one that lets Mike Price catch fish on a Royal Wulff during a Blue-Winged Olive hatch, even though that really shouldn't happen, as far as I can see.

At first I used a good old Zug Bug, then I graduated to the peacock and gray partridge soft hackle I saw in a book by

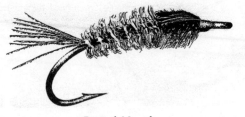

Peacock Nymph

Sylvester Nemes. Then I had a brief flirtation with Prince Nymphs when they started to get popular, but I never could figure out what those two white goose biots hanging off the head were supposed to be, and I finally quit fishing them.

Now I just fish a generic Peacock Nymph. It's tied on a Tiemco 200R in Sizes 14, 16, and 18 with a peacock herl body with a gold wire rib, a peacock thorax, and a wing case of black-dyed goose or duck wing feather fibers. I used to use black-dyed hen-hackle fibers for the tail, and that worked fine, but then I stumbled on some iridescent blue-green peacock body feathers that just add a little more of whatever it is peacock has.

This isn't a fly I fish very often, but it's the one I fish when I don't know what else to do.

Tarcher Nymph

I learned about this fly from its inventor, Ken Iwamasa, and the whole story on it, along with tying directions, is in his book *Iwamasa Flies*, published in 1988. He designed it to copy mayfly nymphs that would "arch their abdomens into a scorpion-like position" when they were dislodged from the bottom and would drift along in the current.

Tarcher Nymph

I liked the nymph because it looked so alive—which is the whole idea behind the pattern—but I also figured out that, because it was a weighted nymph that swam with the hook riding up, it was fairly weedless, so I could swim one over the tops of sunken weed beds or crawl it along the rocky bottoms of lakes and the slow-moving shallows of streams without hanging it up on every cast. Not that it never hangs up, it just doesn't do it on *every* cast.

I've dead-drifted them in current the way Ken intended, and they've worked well—sometimes even a little better than nymphs tied on straight-shanked hooks—but I've come to think of them as fairly specialized bottom flies for trout lakes and for slow backwaters in streams and rivers where trout may take them for mayfly or dragonfly nymphs or maybe small crawdads.

I can picture a certain kind of day when a Tarcher really does it for me. I'm fishing a trout lake, and it's a nothing kind of late morning or early afternoon. There's no hatch, and there are no rising fish at all or maybe just a single lazy dimple every once in a while.

But if I stalk the banks carefully, staying back in the trees or the tall grass so as not to spook the fish, I can spot trout cruising slowly in a few feet of water, usually closer to the bottom than to the surface. Maybe they're opening their mouths now and then to eat a bug, or maybe they're not feeding at all, but they seem sort of restless.

I'll cast a Tarcher Nymph well ahead of a trout and let it settle to the bottom right in his path, or as close to that as I can manage. Fly size depends on the depth of the water. The big ones are heavier than the small ones, so in shallow water I'll use a Size 14 or 16; if it's deeper, I'll use an 8 or 10.

When the trout gets close enough—maybe a foot or two away—I'll give the fly a little scoot along the bottom. I'll decide how much of a scoot by how the fish is acting: If he's moving along pretty quickly, I'll give it a good pull; if he's putting along at dead slow, I'll try just a little twitch. If it works, fine. If not, I'll try something a little bit different on the next fish.

This reminds me of the little bit of bonefishing I've done, except that you don't have a guide perched on a poling platform behind you yelling instructions at the top of his lungs.

As I said, I think of this as a specialized pattern, but it's actually pretty versatile. I've used it for trout in lakes, streams, and beaver ponds and also for largemouth bass, carp, and panfish.

Tarcher Nymphs are tied on English-style bait hooks. I use a Mustad 37160, Size 16 to 8, and I bend it open slightly, just so that the hook point no longer aims right at the eye.

I put the hook in the vise with the hook point down and wrap the shank with lead wire from above the barb to a little bit behind the eye. It's the weight that will roll the fly over in the water so it swims with the hook up. Then I tie in the tail and a piece of copper wire for the rib.

Then I turn the hook upside down in the vise (right side up for the finished fly) and tie a dubbed body from the base of the tail to a little way past the middle of the shank. I tie in a wing case, a hackle feather for the drawn feather legs, and dub the abdomen forward, leaving enough room for the head. Then I fold the hackle feather forward, tie it off, and do the same for the wing case.

Ken's pattern calls for a dark brown or black hackle palmered through the thorax, but I use a drawn brown partridge or Indian hen. The original uses "reddish brown rabbit fur" for the body. I naturally use dark hare's ear dubbing because I have an affection for Hare's Ears that borders on being unhealthy, but you can tie this style of nymph in any color or material you like.

Water Boatman

This is another lake fly. Many of the trout lakes I fish have populations of the aquatic beetles called backswimmers or water boatmen. There are actually two different bugs, but they're so similar that most fly fishers lump them together and use the names interchangeably.

These things are distinctive, and we've all seen them. They're pill shaped, drably colored, small to medium-sized, and have a pair of oversized legs that they use like oars.

They forage on the bottom and breathe underwater by periodically swimming to the surface to get an air bubble that they hold against their abdomens with their short hind legs. They're found in the shallows of lakes and ponds and in the slow backwaters of rivers and streams, not to mention mud puddles and birdbaths.

Boatmen are quick, active, jerky swimmers, and when trout get on them, they can get pretty excited. In shallow

Water Boatman

water you can sometimes see wakes ending in boils or what look like the rolling takes to nymphs just under the surface. In deeper water you can sometimes see fish streaking and turning or just milling around nervously.

I usually see trout feeding on boatmen in the evenings in late summer or early fall. Maybe the bugs are more active then, or maybe the fish just feel safer coming into the shallows when it's getting dark.

The fly I tie is very much like every other boatman pattern you see: a beetle-shaped nymph with a pair of long legs and a shell back. I've tried different things for legs, like rubber hackle and goose biot, but when I learned to tie knotted feather legs for Dave's Hoppers, I settled on that style for the boatmen, too.

Actually, I've seen boatman flies with no legs at all, just a pill-shaped body with a shell back, and they worked just fine.

I use a short, curve-shanked, scud hook—a Mustad 80250 or Tiemco 2457—in Sizes 14, 12, and 10, and I weight the fly with blued bead-chain eyes tied under the hook shank behind the eye. I'll tie in a strip of fibers from a turkey wing or tail-feather for a wing case, dub the back half of the body, tie in the knotted feather legs on each side, pointing backward, then dub the front half of the body, bring the wing case forward, and tie it off right behind the eyes. Then I'll make a fat head with figure-eight wraps of dubbing through the eyes.

I tie a light version with light-colored hare's ear dubbing and a wing case and legs tied from the same mottled-brown wild-turkey wing feather. The dark version uses medium to dark brown dubbing and a wing case of chocolate brown wild-turkey tail with legs from the same feather. I've also seen boatman patterns tied in olive and black.

Some boatman patterns have tinsel or glass-bead bodies to mimic the air bubble, or a strand or two of crinkly tinsel trailing off the back to copy the string of tiny air bubbles these things are supposed to trail behind them as they swim. There are also some tiers who weight their boatmen with heavy lead wraps under the body to make them sink faster.

Everyone seems to have his own ideas about this kind of pattern, but all the flies are fished the same way: Make the cast, let it sink, and then retrieve it in short, jerky strips.

Imitations

Hare's-Ear Nymph—Small

Hook: Tiemco 200R, Sizes 18 to 22.
Thread: Brown 8/0 or finer.
Tail: A small bunch of brown partridge flank feather barbs, tied short.
Body: Dark hare's-ear dubbing, dressed thin.
Rib: Fine copper wire.
Thorax: Dark hare's-mask dubbing.
Wing case: Segment of brown wild-turkey tail or gray goose wing feather.

Hare's-Ear Nymph

Hook: Tiemco 200R for weighted flies, 94840 for unweighted, Sizes 10 to 16.
Thread: Brown 8/0 or finer.
Weight (optional): Lead wire wraps under the thorax.
Tail: A small bunch of brown partridge flank feather barbs, tied short.
Body: Dark hare's-mask dubbing, dressed thin.
Rib: Fine copper wire.
Thorax: Dark hare's-ear dubbing.

Legs: Drawn legs of brown partridge, brown Indian hen, ptarmigan, or
other gamebird.

Wing case: Segment of brown wild-turkey tail or gray goose wing feather.

Hare's-Ear Stonefly Nymph

Hook: Tiemco 200R, Sizes 4, 6, and 8.
Thread: Brown 6/0.
Weight: Lead wraps under thorax and optional finer lead wraps under
forward two-thirds of abdomen.
Tails: Split brown-dyed goose biots.
Body: Dark hare's-ear dubbing, tapered.
Shell back: Brown wild-turkey tail.
Rib: Copper wire.
Thorax: Dark hare's-ear dubbing.
Legs: Drawn legs of brown Hungarian partridge flank feather.
Wing case: Brown wild-turkey tail.

Hare's-Ear Damsel

Hook: Daiichi 1870, Sizes 8 or 10 to 14.
Thread: Brown 8/0.
Eyes: Small blued bead chain or small black Dazl-Eyes.
Tail: Brown partridge flank feather barbs.
Body: Dark hare's-ear dubbing, thin.
Rib: Copper wire.
Thorax: Dark hare's-ear dubbing.
Legs: Drawn brown partridge or Indian hen.
Wing case: Brown wild-turkey tail.
Head: Figure-eight wraps of dubbing through the eyes.

Hare's-Ear Damsel—Olive

Same as above, but with rough, muddy-olive rabbit dubbing, gray
partridge tail and legs, and a wing case of olive-dyed goose wing
feather.

Golden Stone Nymph

Hook: Tiemco 200R, Sizes 4, 6, and 8.
Thread: Yellow 6/0.
Weight: Lead wraps under thorax and optional finer lead wraps under forward two-thirds of abdomen.
Tails: Split golden yellow–dyed goose biots.
Body: Rough golden yellow–dyed rabbit dubbing (or yellow Furry Foam).
Shell back: Brown wild-turkey tail fibers.
Rib: Copper wire.
Thorax: Rough golden yellow–dyed rabbit dubbing (or yellow Furry Foam).
Legs: Drawn golden brown ptarmigan flank or other gamebird feather.
Wing case: Brown wild-turkey tail fibers.

Blue-Winged Olive Nymph

Hook: Mustad 94840,S izes 18 to 24, or Tiemco 200R, Sizes 14 and 16.
Thread: Olive 8/0 or finer.
Tail: A small bunch of mottled or plain gray hen or blue grouse flank.
Body: Pale olive rabbit-fur dubbing.
Thorax: Pale olive rabbit dubbing roughed up to suggest legs (or drawn legs of blue-dun hen hackle on larger sizes).
Wing case: Gray Canada goose wing fibers.

Tricolor Shell-Back Nymph

Same as above, but with a shell back of gray goose and a rib of fine dark olive or brown thread.

Pheasant Tail Nymph

Hook: Tiemco 200R, Sizes 12 to 22.
Thread: Brown 8/0 or finer.
Weight (optional): Lead wraps under thorax.

Tail: Reddish brown cock pheasant tail fibers.

Body: Pheasant tail fibers, wrapped—two to six fibers, depending on hook size.

Rib: Fine copper wire, counter wrapped.

Thorax: Fine peacock herl.

Wing case: Gray goose wing feather segment.

Peacock Nymph

Hook: Tiemco 200R, Sizes 14 to 18 or 20.

Thread: Black 8/0.

Tail: Small bunch of iridescent blue-green peacock body feather barbs or black hen hackle.

Body: Fine peacock herl, dressed thin.

Rib: Fine gold wire, counter wrapped.

Thorax: Fine peacock herl.

Wing case: Segment of black-dyed goose or duck wing feather.

Tarcher Nymph

Hook: Mustad 37160, Sizes 8 to 16. Tied upside down.

Thread: Brown 8/0.

Weight: Lead wraps in the deepest part of the hook bend.

Tail: A small to medium-sized bunch of brown partridge or Indian hen fibers.

Body: Dark hare's ear dubbing.

Rib: Copper wire.

Thorax: Dark hare's ear dubbing.

Hackle: Drawn brown partridge or Indian hen.

Wing case: Brown wild-turkey tail.

Water Boatman

Hook: Mustad 80250 or Tiemco 2457, Sizes 10 to 14.

Eyes: Small blued bead chain tied under the hook shank.

Body: Light hare's-ear dubbing.

Legs: One or two mottled wild-turkey wing fibers, knotted and tied one
    on each side of the body.
Wing case: Mottled wild turkey to match legs.
Head: Figure-eight wraps of light hare's ear dubbing around and ahead of
    eyes.

Water Boatman—Dark

Same as above, but with medium to dark brown dubbing, and tail and
    legs of brown wild-turkey tail fibers.

Chapter 9

Streamers

MY FRIEND CHRIS SCHRANTZ and I don't get to fish together as much as we'd like to. Chris guides out of St. Peter's Fly Shop in Fort Collins, Colorado, so he's pretty busy with paying clients through most of the season. I also do a lot of running around when the fishing is good. When we do get out, it's often to lob streamers from his driftboat on the North Platte River in Wyoming, late in the season when the weather is cool, the days are getting short, and most of the fishermen have gone hunting or just gone home.

For the past few years we've had a running discussion about streamer patterns and fishing. We haven't come to any conclusions, but we've looked at a lot of flies, examined some conflicting theories (and decided they all work), and caught some trout in the process.

I often get more than my share of those trout, only because Chris spends most of the day on the oars. I always offer to take my turn, but he usually says, "No, that's okay." Maybe he really would just as soon row as fish, as he claims, or maybe he's afraid to have me drive his boat.

As for streamer theories, we seem to fall back on tradition about half the time, which is usually a smart thing to do in fishing. Last fall we were planning a float on the North Platte, and Chris suggested I tie up some Platte River Specials. That's not a pattern I've fished much, but Chris said, reasonably enough, that they've been catching fish on them up there for fifty years now and maybe it was time I tried them. I did, and of course they worked.

The original Platte River Special, tied in the early 1950s by Bud Miller of Casper, Wyoming, was a simple feather-wing streamer with four yellow saddle hackles veiled on each side by medium-brown hackles and a wound collar of one yellow and one brown hackle. It's quick, cheap, and easy to tie; it's feathery, fishy, and translucent; and there's something about those colors on that river that just gets it done.

Sticklers for accuracy know that an authentic Platte River Special has no body, just a bare hook shank, but a lot of tiers thought that made the fly look unfinished, so they added a gold tinsel body, and that's how you see a lot of them tied now. Old-timers have been known to turn up their noses at them.

I like streamer fishing because it's not scientific. Or maybe it's just a different kind of science: less stream biology and more fish psychology. There are a lot of realistic-looking sculpin/baitfish sorts of patterns that catch fish, but it's just as likely that something bright and gaudy will work as well or better, and although I prefer simple patterns, sometimes the most complicated, self-indulgent–looking streamers are the ones that work the best.

When you're fishing streamers, you think about the time of year, time of day, brightness of the sun, stream flow, water clarity, the depth your fly swims, where you cast it, how you retrieve it, and so on; but the choice sometimes comes down to tying on either a red and black streamer or a yellow and red one when there's no known reason why a trout should eat either one. Local custom is usually your best guide, but when local custom doesn't pan out, you try something completely different.

Dry-fly and nymph fishing can get you to thinking trout are perfectionist entomologists. Streamer fishing makes you think they're more like vicious predators with poor taste.

I guess my favorite way to fish streamers is from a driftboat on a good-sized river. It's a comfortable, almost luxurious way to fish, and I think one of the keys to streamer fishing is to cover a lot of water, like, say, six or eight miles in a day. I believe that a trout will either eat a streamer right now or never, so if he doesn't pounce on the first or maybe the second cast, it won't do you any good to pound him for another twenty minutes.

When I'm drift fishing, I usually like a weighted streamer on a sink-tip line and a leader no longer than two or three feet to get the whole business down deep quickly. A streamer is fished on a tight line and you could hook a big fish suddenly, so I like to use the heaviest leader I can reasonably get away with. I'll usually tie on the fly with a Duncan loop knot so it doesn't swim as if it were wired to a cable.

I'll often do the same thing with streamers on lakes, either from a boat, from a float tube, or from shore, especially when I'm casting over steep drop-offs and want to get the whole rig down quickly.

In shallower lakes and ponds, small streams, or even bigger rivers in low flows, I'll usually go with a weighted streamer on a floating line with a more conventional leader length of six or eight feet.

That's *usually*, but whenever I think I may end up fishing streamers, I like to have both a floating and a sinking line along.

It's easy to think of streamers as big-water flies, but I've also done well with them on small streams, creeks, and beaver ponds, usually in the late afternoons and evenings when there were few if any trout rising. I'll usually use smaller streamers on small water, figuring that a good fish will be a foot or fourteen inches long, but sometimes I think if I had a little more faith, I might catch bigger fish. Most fertile small trout streams hold a precious handful of real pigs. You just have to get used to the idea of fishing for hours for the chance of a single strike.

Woolly Bugger

According to John Likakis, editor of *Warmwater Fly Fishing* magazine, the Woolly Bugger was invented by Russell Blessing of Lancaster, Pennsylvania, in 1967. It was a combination of a fly with a clipped palmer hackle that Blessing had been working on to imitate a hellgrammite for smallmouth bass, and a streamer called the Blossom Fly that had a marabou tail. The fly was named by Blessing's daughter, Julie.

Woolly Bugger

Woolly Buggers have been tied in every imaginable size and color since then, and they've been used to catch just about everything that swims. It's one of those patterns that tiers can't leave alone, and they're constantly adding tinsel, peacock herl, rubber legs, and so on. Most variations have the word *bugger* in the names, like the Bow River Bugger, but some don't, like the Lectric Leech and the Fire Butt. In parts of Canada they're called Woolly *Boogers*, God knows why.

I had always weighted my Buggers with lead wraps on the hook shank, but when some tiers started using bead chain and dumbbell eyes, I thought that was a good idea. It puts all the weight at the front so the fly jigs nicely in the water, and it lets you tie a trimmer body. I've never been able to decide what Buggers are supposed to be, but I somehow got it into my head that they should be tied sparsely.

I used to paint my lead eyeballs, but when the already painted ones started going on sale, I switched to those just to save time. Yellow eyes with black dots seem to be good, all-round, highly visible colors.

There's a theory that predatory gamefish key in on the eyes of their prey, and it might be true, because no less a fisherman than Left Kreh says so, and because tiers have been putting eyes on streamers for a long time. Some of the old feather-winged streamers have jungle cock nails for eyes, sometimes tied over flank-feather cheeks to represent a head. Other tiers have painted eyes onto the lacquered thread heads of streamers; and, more recently, teddy bear and taxidermy eyes have been glued onto wool and clipped deer-hair streamer heads.

As I said, there could be something to it, although I've caught plenty of fish on streamers with all the paint chipped off the eyeballs.

I tied my Buggers on Tiemco 200R hooks for a long time but switched to Daiichi 1870 Swimming Larva hooks as soon as they came out. The bowed shank of the hook makes the fly look more lively, and I like the way it swims with the hook down, even with lead eyes on top.

I tie my Buggers mostly in black and dirty olive in Sizes 4 to 10. I first lash on the eyes, tie in a sparse marabou tail just above the hook barb, then lash the butt of the feather forward along the hook shank to the eyes to make a smooth underbody.

I tie the saddle hackle in by the tip right at the base of the tail, and then dub a sparse body of rabbit fur forward, tapering slightly toward the eyes. Next, I wrap the hackle forward in fairly loose turns, tie it off behind the eyes, and make a head by wrapping figure eights of dubbing through the eyes.

Tying the hackle in at the tip makes the barbs lie back like wet-fly hackle and gives you shorter barbs in the rear that taper to slightly longer ones in front, to go with the gradual taper of the dubbed body ending in a fatter head. It's just the way I like them to look.

I like to fish Woolly Buggers in lakes, especially early in the season or when the water is off color, but I'm likely to fish one anytime, anywhere, fast or slow, deep or shallow, day or night. Apparently, I'm not alone in that. If you're stumped and wonder aloud what to do next, at least half the fly fishers in North America will say, "I don't know, try a Bugger."

Muddler Minnow

I've tied Muddlers for years, even though spinning what I thought were good deer-hair heads was always a struggle for me. Muddlers made by good tiers had heads spun as tight and

Weighted Muddler

perfect as cork, but mine always had the consistency of a pile of kindling wood. Trout would eat them often enough—and a trout is the ultimate judge of a fly—but I was just never proud of them.

The first time A.K. and I went to the Minipi River drainage in Labrador, Canada, to fish for those giant brook trout, I tied up a couple dozen Muddlers. Actually, I sweat blood over them for weeks, spinning heads from the fluffiest deer hair I could find and packing them so tightly with one of those slotted brass packing tools that my fingers ached. I thought they looked pretty good.

During shore lunch on the second or third day in Labrador, I took them out and showed them to our guide, Ray, who said, "These are no good, ey."

"Why the hell not?"

"Well, because Muddler heads should be spun loose so you can squeeze 'em wet so they don't float."

"Oh."

The original Muddler Minnow was tied by Don Gapen in Canada in 1937 and later popularized by Dan Bailey in Montana. (The oldest Muddlers I've seen do, in fact, have pretty loosely spun heads. I just thought they didn't know how to do it right back then.) Since that time the Muddler Minnow has been redesigned into the Marabou Muddler, the Missoulian Spook, the Spuddler, and a dozen or so others, and I think it's fair to say that all of the deer-hair- and wool-headed sculpin patterns you see now are direct descendants of the Muddler.

The deer-hair heads on many old Muddlers were fat and round, and the deer-hair collars extended all the way back to the point of the hook. Later Muddlers often have more tightly packed heads clipped to a flatter, more sculpin-like shape, and shorter collars; but other than those minor stylistic changes, the original pattern hasn't changed much in more than sixty years, and it still catches trout.

I used to weight my Muddlers by wrapping lead on the back half of the hook shank and covering it with crinkly tinsel, but as soon as lead dumbbell eyes came out, I started trying to figure out how to use them on Muddlers. (At this stage of the game, I probably couldn't fish streamers at all without dumbbell eyes.) Then, when I discovered the Daiichi Swimming Larva hook that would make a lead-eyed streamer swim right side up, it seemed obvious if I could just put it all together.

You actually can spin and stack a deer-hair head around a pair of eyeballs, but it's clumsy, laborious work, and I could never pull it off more than two out of three times.

A.K. inadvertently solved the problem for me. I was tying some wool-headed streamers, and I told him I was having trouble with them. He said it was easy, just spin the wool on a dubbing loop as if you were making a thick hair hackle, wrap it on, and trim it. Naturally, it worked (every tying tip A.K. has ever given me has worked), and in a stroke of something resembling genius, I tried it with deer hair.

That worked, too. You spin a generous bunch of deer hair with stacked tips into a dubbing loop and wrap it on ahead of the body, combing the tips back as you go, to make the collar. Then you spin a second bunch clipped short for the head. The dubbing loop spins the hair into a kind of bottle brush, so you have to comb it back with your fingers with every turn, but with a little practice it goes on smoothly. Do a figure-eight wrap through the lead eyes, and make one or two more wraps ahead of them. Tie off the loop and trim the head.

This doesn't take me any longer than stacking and packing deer hair, and I get a more uniform collar and a neat, but loosely spun, head. It's also the only way I can think of to spin deer hair around a set of eyes.

I tie these on Daiichi 1870 hooks in Sizes 4, 6, 8, and sometimes 10, and except for the painted dumbbell eyes and the newfangled hook, they're traditional Muddlers. The tail is made from a matched pair of mottled turkey wing sections; the body is flat, gold tinsel; the underwing is brown bucktail; the wing itself is paired mottled turkey to match the tail; the collar is stacked deer hair, and the head is spun and clipped deer hair.

I like Muddler wings to lie low along the body instead of sticking up at a steeper angle, so I'll tie each wing in separately on the sides of the shank, like the quill wings on an Atlantic salmon fly, instead of squashing them together on top.

I think of the Muddler as an all-round, naturally colored western freestone pattern. I like a Size 4 from a driftboat in big water and something like an 8 or 10 on a floating line for small streams and pocket water.

Bunny Flies

I've been fooling around with streamers tied from strips of tanned, hair-on rabbit hide—what we now call Zonker strips—since I first saw them in the 1980s, and apparently so has just

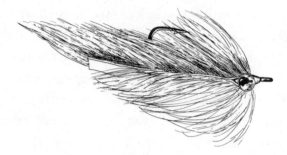

Bunny Streamer

about everyone else. There's Dan Byford's original Zonker, Mike Lawson's Wool Head Sculpin, Larry Dahlberg's Rabbit Strip Divers, and Dave Whitlock's Waterpups, Waterdogs, and Watersnakes, just to name a few that you see tied commercially.

I've also seen an odd assortment of Matuka-style streamers and leech patterns tied with rabbit strips and some tiers like to use them as substitutes for saddle hackles on feather wing patterns.

The first rabbit-strip pattern I ever saw was a fly tied by Sandy Pittendrigh called a Roadkill Streamer. It had a strip of natural, brownish gray rabbit lashed to an otherwise bare hook with a silver propeller and a couple of brass beads at the head. It didn't look like much, but it was as simple and elegant in its way as a worm on a hook.

The first one I tied and used was a thing called a Bunny Fly by the northern pike fishermen at Spinney Mountain Reservoir in Colorado. It had a long rabbit strip trailing off the back of a short-shanked hook, with another strip wound on as a collar hackle ahead of that, sometimes with bead-chain eyes for weight. It was usually tied in all black, and with tail and hackle combinations like black and red, white and red, and yellow and red.

Then I got interested in Scott Sanchez's Double Bunnies because they looked fuller and more minnowlike, and also because they were more streamlined and easier to cast.

A Double Bunny has two rabbit strips laid along the top and bottom of a hook and glued with the skin sides together over the weighted shank. Naturally, you have to thread the hook through one strip, which you do by putting a hole in it with a leather punch, making a slit with a razor blade, or just poking the hook through the skin and pulling it on.

Sanchez's original pattern had a snelled trailer hook behind the fly, tinsel trailing down the sides, and glued-on eyes. Most tiers did away with the trailing hook, and some didn't bother with the eyes or the tinsel, either. The only rule was, if you were using two colors of rabbit, the darker one was always on top to match the universal color scheme of fish.

Not long after that I saw my first Whitlock Hare Grubs and Hare Jigs. These used the same construction as the Double Bunnies except that they had lead eyeballs tied on the top of the hook, which meant that the fly rode with the hook up and the darker strip was now on the hook bend side of the shank. These were tied as bass flies, so the top strip could be as long as seven inches, but the bottom strip didn't extend much past the hook bend. Double Bunnies were sometimes also tied that long, but on that pattern the two rabbit strips were the same length.

I liked Whitlock's flies because they were as easy to tie as Double Bunnies, but they rode with the hook up, so they were less likely to snag bottom. At first I tied them on No. 2, 4, and 6 stinger hooks and fished them for bass and pike in colors like red and yellow, chartreuse and white, black and red, and all black or all purple.

They worked so well on warmwater fish that I started to tie them on smaller-gapped, ring-eyed streamer hooks for trout: the same flies except with shorter tails. I also used some more sedate colors, such as natural gray and white, brown and white, beige and white, black and white, yellow and white, all white, and either all olive, olive and yellow, or olive and orange.

The Whitlock flies have tinsel and multicolored rubber legs dangling out of the bodies, but I did away with that stuff after a while and didn't notice any difference. On the other hand, I've tried some fancy stuff of my own, like wound collars of schlappen hackle, spun-hair hackles, and saddle-hackle veiling. (I like the ease and simplicity of the fly, but I'm a tinkerer at heart and I just couldn't help myself.) But I finally settled on the stripped-down workhorse model with just the two rabbit strips, dumb bell eyes (either painted or silver), and a thread head that more or less matches the darker color of rabbit.

I cover the hook shank with tying thread (to give the glue something to bite) and wrap the eyes on first. I put them on what would normally be the top of the hook shank so that the fly will swim upside down, and I leave plenty of room between the lead eyes and the eye of the hook to tie down the rabbit strips.

Once the eyes are lashed on, I'll take the hook out of the vise and thread my darker rabbit strip through the point of the hook and position the strip along the shank before I put the hook in the vise, with the bend down. Then I'll tie that strip in behind the hook eye, tie in the lighter strip on the other side, and finish the head. At this point I'll trim the lighter piece so that the skin comes just to the hook bend, and trim the longer, darker piece so that it extends at least one hook-shank length past the bend.

(Actually, I'll often leave that top piece *real* long—as much as two or three inches past the bend—figuring I can always clip it short later without hurting the looks of the fly.)

Finally, I'll cover the skin sides of both strips with glue where they come together (don't add glue to the end of the longer strip), and carefully squeeze them together over the shank.

There's been some discussion about what kind of glue to use for this kind of fly, but I've had good luck with regular old rubber cement from an office supply store. So far, I've never had one come apart, and that includes flies that were chewed on by things like pike and bull trout.

Wool-Head Bunny Streamer

These two-strip rabbit streamers are big, fat flies by nature, especially when you tie them with magnum-sized Zonker strips like I usually do, and they're great for big, hungry fish in fast or roily water, but sometimes they're a little too big.

Lately, I've been tying a wool-headed, sculpin/Matuka version that I'm really starting to like. It still has the substantial quality of a Zonker strip streamer, but it's a little trimmer

Wool-Head Bunny Streamer

and a little more understated. I tie it on the Daiichi 1870 hook with a Matuka-style bunny wing over a dubbed body that either roughly matches or complements the wing color, a long beard of white bucktail extending back to the hook bend underneath, small lead eyeballs, and a collar and clipped head spun from the same color rabbit fur as the wing. I tie these in what I think of as the standard streamer colors: gray, brown, beige, dirty gold, olive, black, yellow, or white. (Chris Schrantz once said that white is an underrated color for streamers, and I think he's right.)

Earlier I mentioned conflicting streamer theories that all seemed to work. I really do believe in that. A sparsely dressed Woolly Bugger is just the merest suggestion of something alive swimming through the water; a Muddler is a fat-headed, thin-bodied, sculpin sort of thing; and a Bunny Fly is a big, juicy piece of meat.

Mud Bug

What I call a Mud Bug (southern slang for crawdad) turns out to be a slight variation of Bob Clouser's great Foxee Red

Mud Bug

Clouser Deep Minnow, although I didn't know that when I first tied it.

Clousers are tied on straight-eyed streamer hooks with lead eyes on top of the shank so that, once again, the fly runs deep and swims with the hook bend up. Clousers went almost immediately from being a series of specific patterns to a general tying style, and in the years since they first came out I've seen them tied in just about every color combination you can think of with either natural hair or synthetics. Most are still simple, thin bucktails, light on the bottom and dark on top, and belong to the merest-suggestion-with-goggle-eyes school of streamer design.

Anyway, I'd been looking for a good crawdad pattern but couldn't find anything I liked. Most crawdad flies seem to be tied as much for fishermen to admire as to catch fish. They're realistic and beautiful and if you were a fish you'd eat one, but one of the secrets of the sport is that fish aren't like us. These flies were also too difficult and too time consuming to tie, and that's not good for a fly that's meant to bump along the bottom amid rocks, weeds, stumps, and rubble.

I've tied crawdad patterns with shovel-shaped tails, carefully segmented bodies, legs, whiskers, and pinchers tied from cut feathers or clumps of hair. They've worked okay, but it seems like the fancier they were and the longer they took me to tie, the more likely I was to wedge them between two boulders on the first cast and have to break them off.

I also got to watching crawdads in trout lakes and decided that most patterns were way more detailed than they had to be to catch fish, and possibly too visible, as well. The crawdads I saw were so close to the mud color of the bottom that I didn't see them unless they moved. And when they moved, all I could make out were indistinct, muddy blurs on the bottom. I decided that the trout feeding on these things probably didn't see much more than I did: just a size and a movement and maybe a puff of silt. The fish wouldn't be thinking "crawdad." In their single-mindedly predatory way, they'd be registering "camouflaged, moving (and therefore edible) thing."

One day I tried twitching a plain brown bucktail version of a Clouser Minnow along the brown silt bottom of a lake in a few feet of water near shore, and I caught as many trout as I'd ever gotten on fancier crawdad flies. Unable to leave well enough alone, I tied a Clouser with crinkly brown bucktail underneath and a short but overly bushy wing of red fox body fur with lots of grayish underfur and the reddish guard hairs trailing out behind. It also had the red-painted lead eyes with black pupils that are a trademark on most Clouser patterns.

The first batch was tied on Size 4 and 6 Tiemco 200R hooks, and I used pretty large eyes to get the fly down to the bottom quickly and keep it there. They worked so well that I burned through them in a month (as you tend to do with bottom flies), and on the next batch, thinking "crawdad," I added brown rubber legs.

Sometime later, I was thumbing through a catalog of Umpqua Feather Merchants fly patterns and came upon the Foxee Red Clouser by Bob Clouser himself. It's tied with red fox tail top and bottom instead of the bucktail and fox body I use, it lacks rubber legs and has some tinsel, but it's close enough to prove, once again, that most of the good ideas in fly tying already existed by the time I got there.

Those are the streamers I depend on the most, but I fish plenty of others. I'm almost always willing to try something new, and I always keep an eye out for local patterns on new rivers.

The only time I ever fished a Bow River Bugger was on the Bow River, where I got a brown trout big enough to get my picture (posing with the fish and my guide, Dave Brown) in a Calgary newspaper. Now I carry a couple of big ones, just in case I'm ever in a driftboat with a reporter again.

I have only one Platte River Special left over from that last float on the North Platte, but I'll have at least a half dozen more in a couple of sizes before I go back. The ones I tied had weighted shanks with crinkly gold tinsel bodies, but I really should try some with the original bare shank because it could make a difference. And while I'm at it, I should tie some of Chris's neat marabou Spey versions of the same pattern.

And so it goes.

Imitations

Woolly Bugger

Hook: Daiichi 1870, Sizes 4 to 10.
Thread: Black 6/0.
Eyes: Lead dumbbell eyes, painted yellow with black pupils, tied on top of the shank in the dip behind the hook eye. Medium to extra small eyes, depending on hook size.
Tail: Black marabou blood feather.
Body: Black rabbit-fur dubbing, tied thin, tapering toward the front.
Hackle: Black saddle hackle tied in at the tip and palmered forward.

Head: Figure-eight wraps of black rabbit dubbing through and ahead of they eyes.

(This fly is also tied in olive, brown, or any other color you like with matching thread, tail, dubbing, and hackle.)

Weighted Muddler

Hook: Daiichi 1870, Sizes 4 to 10.

Thread: Tan 6/0.

Eyes: Lead dumbbell eyes, painted yellow with black pupils, sizes medium to extra small, depending on hook size.

Tail: Matched pair of mottled-oak wild-turkey wing segments tied in with the curve down.

Body: Flat gold tinsel.

Underwing: Brown bucktail extending to hook bend.

Wing: Matched pair of mottled oak wild-turkey wing segments tied in tent style with one segment on each side of the hook shank, extending to the end of the tail.

Collar: Stacked natural deer hair spun on a dubbing loop and wrapped hair-hackle style.

Head: Natural deer hair spun on a dubbing loop and wrapped hair-hackle style behind, through, and ahead of the eyes and trimmed to shape.

Bunny Streamer

Hook: Mustad 9674, Daiichi 1750, Tiemco 9395, or other comparable, straight-eyed streamer hook Size 2, 4, or 6.

Thread: 6/0, color to match darker rabbit strip.

Eyes: Lead dumbbell eyes, either painted or silver, lashed on top of the hook shank to make the fly swim upside down.

Underbody: Thread wraps along the length of the hook shank to hold the glue.

Wing: Two rabbit strips, one longer and darker in color than the other, tied in ahead of the eyes and glued together over the hook shank. The longer, darker strip is on the hook bend side, with the hook point threaded through the skin, and rides on top of the fly in the water. Both strips are tied in ahead of the eyes.

Wool-Head Bunny Streamer

Hook: Daiichi 1870, Sizes 4 to 8.
Thread: 6/0, color to match rabbit strip.
Eyes: Lead dumbbell eyes, painted or silver, tied on top of the hook shank in the dip behind the hook eye.
Body: Dubbing to match (or complement) the rabbit strip.
Wing: Zonker strip tied in behind the eyes and extending to, or just past, the hook bend.
Rib: Fine oval gold or silver tinsel, Matuka style.
Beard: White bucktail extending back to the hook bend.
Collar: Rabbit fur to match the wing, spun on a dubbing loop and wrapped hair-hackle style.
Head: Rabbit fur to match the wing, spun on a dubbing loop, wrapped hair-hackle style behind, through, and ahead of eyes, and trimmed to a flat-headed sculpin shape.

Mud Bug

Hook: Tiemco 200R, Sizes 4 to 8.
Thread: 6/0 brown or tan.
Eyes: Lead dumbbell eyes, painted red with black pupils, tied on the top of the hook shank to make the fly swim upside down.
Legs: Brown rubber legs lashed to the hook shank and separated with figure-eight wraps of tan dubbing.
Bottom wing: Brown bucktail tied in ahead of the eyes with the butts covered with thread back to about the forward third of the hook shank.
Top wing: A large bunch of red fox body hair with lots of gray underfur, with the reddish brown tips extending past the hook bend.